A TEXTBOOK
OF
AGRICULTURAL
ENTOMOLOGY

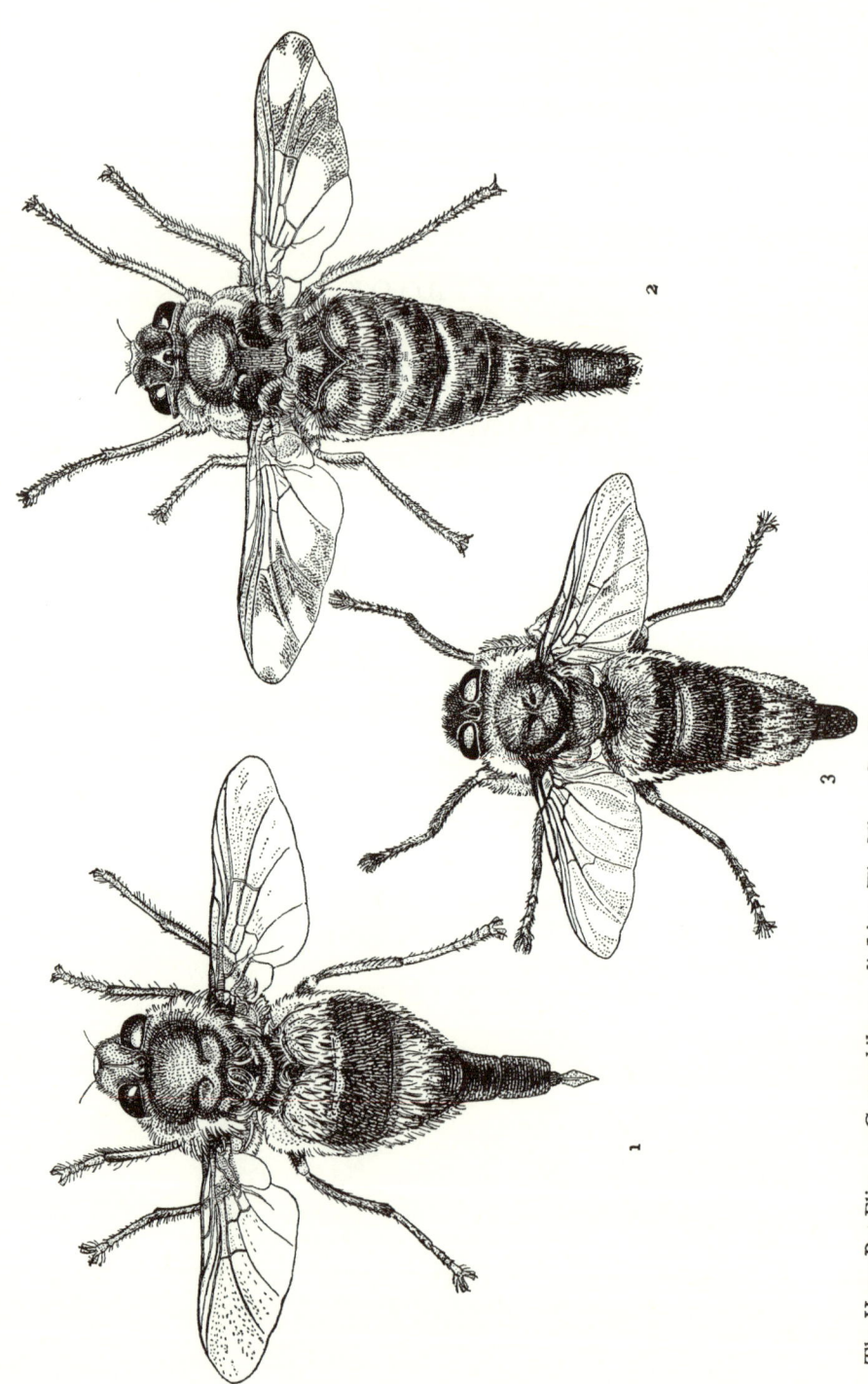

The Horse Bot Flies. 1. *Gastrophilus nasalis* Linn. The Throat Bot Fly. 2. *Gastrophilus intestinalis* de G. (*equi* F.) The Common Bot Fly. 3. *Gastrophilus haemorrhoidalis* Linn. The Nose Bot Fly. All females. × 3 approx. (See p. 192.) (Redrawn from Hadwen and Cameron.)

A TEXTBOOK
OF
AGRICULTURAL
ENTOMOLOGY

by

KENNETH M. SMITH
D.Sc., Ph.D., F.R.S.
Director, Plant Virus Research Station, Cambridge

CAMBRIDGE
AT THE UNIVERSITY PRESS
1948

CAMBRIDGE UNIVERSITY PRESS
Cambridge, New York, Melbourne, Madrid, Cape Town,
Singapore, São Paulo, Delhi, Mexico City

Cambridge University Press
The Edinburgh Building, Cambridge CB2 8RU, UK

Published in the United States of America by Cambridge University Press, New York

www.cambridge.org
Information on this title: www.cambridge.org/9781107661028

First edition 1931
Second edition 1948
First published 1948
First paperback edition 2013

A catalogue record for this publication is available from the British Library

ISBN 978-1-107-66102-8 Paperback

To

My Wife

CONTENTS

PREFACE TO SECOND EDITION

Possibly the development most significant for agricultural entomology which has taken place since this book was first published
is the preparation of the new insecticides D.D.T. and Gammexane.
Since the application of these substances to the control of insect
pests of agricultural crops has as yet hardly begun it is difficult to
say how successful they will be. So far as they have been tried they
seem to show great promise and I have endeavoured to indicate
each case where success has been achieved.

The bulk of the book remains unchanged but any new facts
concerning life-histories and control methods have been incor-
porated so far as possible.

The study of the insect-borne virus diseases affecting agricultural
crops has made great strides and in consequence Chapter XIV has
been entirely re-written and some new illustrations added. I am
indebted to my colleague Dr Roy Markham for taking the photo-
graphs for Figs. 82, 84.

KENNETH M. SMITH

CAMBRIDGE

PREFACE

A book of the nature of the present volume must necessarily be
something of a compilation, and in preparing it I have made free
use of the many papers on entomology which have appeared in the
various scientific journals. I have, however, endeavoured to make
acknowledgment of the source of any new or important facts and
observations which I may have quoted. In surveying this literature
the invaluable aid rendered by the *Review of Applied Entomology*
must be mentioned.

Special acknowledgment is due to my sister, Faith Blomfield
Smith, for her care and skill in reproducing many of the figures
borrowed from other sources and for making the original illustrations of Figs. 13–18, 20–22, 40 and 43. I am also much indebted
to Miss M. E. Sewell for preparing Fig. 37, and for her help
in other ways; to Miss Sofie Rostrup for the loan of the original
drawings of Figs. 28–30, and for permission to make use of others;

to Dr H. F. Barnes for some valuable information upon the Cecidomyidae; to Mr J. P. Doncaster for preparing Fig. 8; to Dr A. D. Imms, F.R.S., for reading part of the manuscript, for his advice on the literature of the subject and for the loan of two original drawings; to Dr George Salt and Mr H. C. F. Newton for some unpublished records of parasites of *Cephus pygmaeus* and *Phyllotreta* spp. respectively; and to Dr W. R. Thompson, of the Imperial Institute of Entomology, for the facilities afforded to me at the Farnham House Laboratory, whereby I was enabled to obtain much information in regard to the natural enemies of many insect pests. Acknowledgment for permission to reproduce figures is also due to Miss D. J. Jackson, Dr F. C. Bishopp, Dr A. E. Cameron, Dr N. Criddle, Dr J. Davidson, Dr S. Hadwen, Mr W. E. H. Hodson, Dr N. A. Kemner, Mr J. Keys, Dr O. Lundblad, Dr J. N. Oldham, Mr T. H. Taylor and the University of Leeds, and the late Mr F. V. Theobald. In every case acknowledgment is made to the author beneath each borrowed illustration. In addition, blocks were kindly provided for Fig. 41 by the Royal Physical Society of Edinburgh; for Figs. 61, 68 and 69 by Messrs Benn Bros. from *Insect Pests of the Horticulturalist*; for Figs. 6, 9, 10, 25 and 26, by Sir G. A. K. Marshall, F.R.S., from the *Bulletin of Entomological Research*; and Figs. 1, 3, 23, 34, 36, 75, 78, 79 are reproduced by permission of the Controller, H.M. Stationery Office, and the Ministry of Agriculture and Fisheries. The Editors of the *Annals of Applied Biology* allowed the use of blocks for Figs. 35, 64–66, 76 and 77; Messrs Methuen supplied electros of Figs. 32 and 33, from *A General Textbook of Entomology*; and the Trustees of the British Museum allowed the use of Fig. 53 from the *Second Report on Economic Zoology*. Mr H. Britten took the photograph reproduced in Fig. 69 and Mr C. W. Williamson took those reproduced in Figs. 78 and 79.

I am also indebted to the Chief of the Division of Publications, U.S. Department of Agriculture, for permission to reproduce a number of illustrations, and to Professor J. Stanley Gardiner, F.R.S., and Mr L. E. S. Eastham for permission to work at the insect collections in the Zoological Museums at Cambridge.

KENNETH M. SMITH

SCHOOL OF AGRICULTURE
UNIVERSITY OF CAMBRIDGE
December, 1930

LIST OF ILLUSTRATIONS

CHAPTER I

INTRODUCTORY

The need for an up-to-date textbook on agricultural entomology has been apparent for many years. Although there are now in existence one or two modern books dealing with the insect pests of farm crops in Great Britain, yet these are elementary in character and from their nature are unable to deal as fully with the subject as could be wished in view of the valuable information on applied entomology which has accumulated during the last two decades. Curtis' *Farm Insects* published in 1860 remains the standard textbook on agricultural entomology in the British Isles but this is now long out of date.

Although under present conditions of high cost of production, the publication of as large and profusely illustrated a volume as Curtis' *Farm Insects* would be a matter of great difficulty, yet it is this book which the writer has had in mind in the preparation of the present volume. An attempt has been made to give as complete and up-to-date an account of each insect dealt with as the limitations of space would allow. The descriptions of the adult insect and its various stages are given in some detail, together with the salient facts in its life history. Special attention has been paid to control methods, and though descriptions must necessarily be brief, the writer has endeavoured to give essentials. In each case the more important farm weeds which act as alternate hosts for many insects are given and also the natural enemies so far as these are known. The book is based upon the results of much original work, and it is hoped that the information thus assembled and the manner of its presentation will render it of use to two classes of readers, both the agriculturist and the agricultural entomologist. The writer has omitted any introductory account of the elements of entomology as there are already a number of excellent books on the subject, to which references will be found at the end of this chapter. No attempt has been made to deal with the insect pests of fruit, partly because they could not adequately be described in a volume of this size and partly because there already exist two comprehensive books,

F. V. Theobald's *Insect Pests of Fruit*, and A. M. Massee's *Pests of Fruits and Hops*. Insecticides play a comparatively small part in the control of insect pests of the farm and in consequence there is no separate chapter dedicated to their description and preparation. Where it is considered that sprays or dusts would be of use, instructions for their preparation are given in the section on control for the insect concerned.

It will not, perhaps, be out of place in a book of this nature to give a short description of the organisation of agricultural entomology in England and Wales at the present time. Since this book was first published the Phytopathological Service, which included entomology and mycology, has been replaced by the National Agricultural Advisory Service which changes the status of the Advisory Entomologist from county to national employment. The new features of this national service include the setting up of Provisional Centres to provide specialist advice to the counties within the province (see Fig. 1), increased numbers of specialist officers and in due course the establishment of experimental farms where new products, machinery or methods can be tried out under practical conditions.

The activities of the National Agricultural Advisory Service may be set out under three heads: first, there is the individual problem or difficulty which crops up from time to time on the particular holding and which must be dealt with as it arises. Secondly, there is a large and rapidly growing body of scientific and technical knowledge that bears upon the business of farming and food production and which must be conveyed, in practical terms, to food producers. Thirdly, there are the many questions that may fairly be asked by practical men but to which science has at present no answer; here the advisory officer must either set on foot his own investigations or, if his facilities are unequal to the task, must state the problem to the research workers and invite their help. (J. A. S. Watson (1946), *Agric. Lond.*, LIII, 375–380). The following is a list of some of these research institutes: Rothamsted Experimental Station at Harpenden; East Malling, Kent; Long Ashton, Bristol; Cheshunt in the Lea Valley; the Imperial College of Science and Technology, London; and a number of stations attached to the School of Agriculture, Cambridge. As regards the Plant Pathological Laboratory at Harpenden, among its main duties are the provision

of a scientific basis for the Orders issued under the Destructive
Insects and Pests Acts and the collection and distribution of
information regarding the spread of pests and diseases.

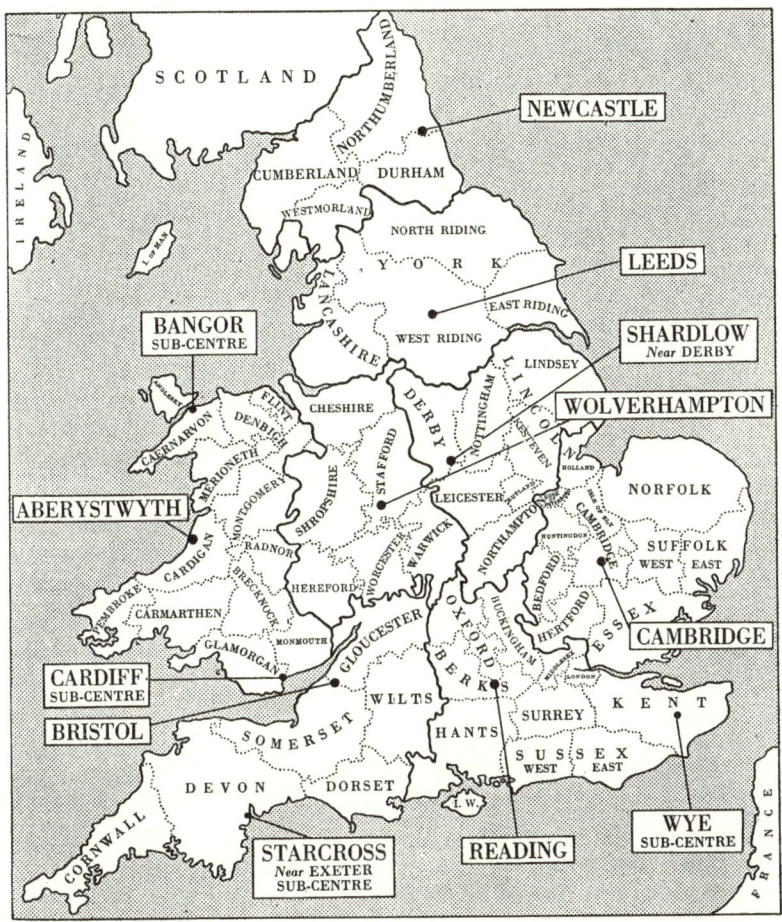

(From J A. S. Watson, *Agriculture*, London, 1946.)
Fig. 1.

In addition to the above organisations, there exists the Imperial
Institute of Entomology, the headquarters of which are at the
British Museum (Natural History). This was founded in 1913 for

the purpose of encouraging and co-ordinating entomological work throughout the British Empire, in relation to agriculture and also to human and animal diseases. The Institute publishes a quarterly journal, the *Bulletin of Entomological Research*, and a monthly journal, *The Review of Applied Entomology*. The latter appears in two parts: (A) Agricultural: (B) Medical and Veterinary and summarises the current entomological literature. This is an invaluable publication to the agricultural entomologist and enables him to keep in touch with entomological progress in all parts of the world.

REFERENCES

GENERAL ENTOMOLOGY

Imms, A. D. (1945). *Outlines of Entomology*. 2nd edn. Methuen and Co., Ltd , London.

Advanced.

Imms, A. D. (1946). *A General Textbook of Entomology*. 6th edn. Methuen and Co., Ltd., London.

GENERAL WORKS ON INSECT PESTS

Elementary.

Ormerod, E. A. (1890). *A Manual of Injurious Insects*. Simpkin, Marshall, Hamilton and Kent, London.

Theobald, F. V. (1899). *A Textbook of Agricultural Zoology*. W. Blackwood and Sons.

Petherbridge, F. R. (1923). *Fungoid and Insect Pests*. Cambridge University Press.

Fryer, J. C. F. and Brooks, F. T. (1928). *Insect and Fungus Pests of the Farm*. Ernest Benn.

Robinson, D. H. and Jary, S. G. (1929). *Agricultural Entomology*. Duckworth, London.

Collected Leaflets on Insect Pests of Farm and Garden Crops, 1928. Sectional Volume, No. 11. Ministry of Agriculture and Fisheries, London.

Advanced.

Curtis, J. (1860). *Farm Insects*. Blackie and Son.

Sorauer, Paul. *Handbuch der Pflanzenkrankheiten*, Fünfter Band. Tierische Schädlinge an Nutzpflanzen, II. Teil. Berlin, 1928.

JOURNALS

General Literature.

Review of Applied Entomology, A and B. Imperial Institute of Entomology, 41 Queen's Gate, London, S.W. 7.

Current Work.

Annals of Applied Biology. Cambridge University Press.

Bulletin of Entomological Research. Imperial Institute of Entomology, London, S.W. 7.

There are occasionally entomological papers of practical importance in the following journals:

Agriculture. (London.)

Welsh Journal of Agriculture. (University of Wales Press Board. Cardiff.)

Scottish Journal of Agriculture. (Edinburgh.)

Transactions of the Highland Agricultural Society. (Edinburgh, William Blackwood and Sons.)

GENERAL WORKS ON INSECT CONTROL

BOURCART, E. (1926). *Insecticides, Fungicides and Weedkillers.* 2nd English edn. Ernest Benn, London.

MARTIN, HUBERT (1940). *The Scientific Principles of Plant Protection.* 3rd edn. London, Edward Arnold.

WARDLE, R. A. and BUCKLE, P. (1923). *The Principles of Insect Control.* Manchester University Press.

WARDLE, R. A. (1929). *The Problems of Applied Entomology.* Manchester University Press.

CHAPTER II

METHODS OF INSECT CONTROL AND THEIR APPLICATION IN FARMING PRACTICE

Modern insect control may be approached from four main standpoints and these will be briefly reviewed in their application to the insect enemies of agricultural crops. There is, firstly, the direct application of chemicals, otherwise insecticides, to kill or deter the insect; secondly, there are what may be termed 'cultural methods', where the agriculturist by variation in his farm practice endeavours to put the insect continually at a disadvantage. The third method, known as 'biological' or 'natural' control, invokes the aid of parasites and predators which are artificially introduced into the neighbourhood of the pest in question. Lastly, the importation of foreign insects and the movement of resident pests within a country are restricted by means of legislation.

CHEMICALS. The use of insecticides in the agricultural practice of the British Isles is very limited and is governed largely by the question of price, the profit on the average farm crop, especially at the present time, allowing small margin for expenditure on insecticides. The use of chemicals and sprays, however, has been found practical and efficacious in the case of certain insects and some of these instances are briefly indicated. The most recent development is the control of leatherjackets (*Tipula* spp.) and cutworms (*Euxoa, Feltia*, etc.), by the use of Paris green (acetoarsenite of copper) as a poison bait. (See pp. 74, 161.) This has proved a cheap and practical method of control for these two serious agricultural pests. The use of calcium cyanide as a soil insecticide is practicable under certain conditions and though too expensive for application on a large scale, it may be used for poisoning wireworms which have previously been assembled in a small area by the use of bait crops. (See p. 108.) Crude naphthalene has also proved its worth as a cheap soil insecticide in certain cases. As regards vegetables some success has been achieved with the use of tar oils as deterrents and of mercury compounds as poisons against

Dipterous pests. (See p. 221.) Certain chemicals have been proved efficacious for killing the larvae of the warble fly in the backs of cattle; among these may be mentioned derris-iodoform and vaseline and a mixture of nicotine sulphate, calcium hydrate and water. (See p. 203.)

CULTURAL CONTROL. The farm operations which can be utilised or manœuvred to place potential or actual insect pests at a disadvantage are mainly the following: (a) time of sowing; (b) crop rotation; (c) cultivation of the ground; (d) manuring; (e) harvesting; to which may be added (f) farm hygiene.

(a) *Time of sowing.* By timing the growth of a crop so that the vulnerable vegetative period does not coincide with the maximum numbers of the insect pest, much damage may be avoided. Early sowing is generally an advantage with spring-sown crops in order that the young plants may have become established before the appearance of the insect pests; a good example of this is the early sowing of spring oats to avoid the frit fly. It is important, however, that such early sowing be done with due consideration of the local climatic conditions, otherwise the plants may not have made the necessary progress at the time of appearance of the insect and the desired effect will not be attained. In districts where oats suffer seriously from attacks by leatherjackets, early sowing is not always an advantage. In this case early sowing lengthens the period before the formation of the adventitious roots, a period when the oat is particularly susceptible to attacks by the larvae, and so renders the oat liable to attack for a longer time than if sown later. In districts where the carrot fly (*Psila rosae*) is prevalent, it has been found advantageous to sow carrots later than the usual practice; by this means the growth is so arranged that the carrots are not above ground until the majority of the first brood of carrot flies have disappeared and the number of potential enemies is thus appreciably reduced.

(b) *Crop rotation.* By variation in the rotation and arrangement of his crops the farmer endeavours to avoid growing a crop in or near to a field which has been infected with a certain pest the previous year. An example of the benefit accruing from the careful selection of crops is seen in the case of *Hylemyia coarctata*, the wheat bulb fly. This insect has the habit of ovipositing in the bare soil, particularly of fields carrying potatoes or in bare fallow, and therefore wheat, which is a susceptible crop, should follow grass or beans or some crop which covers the surface of the ground, rather than potatoes or bare fallow.

Although tending to reduce local loss from insect damage, Wardle (9) considers that crop rotation in practice probably has little influence upon the degree of infestation over the whole area, chiefly because of the variation in crop rotation on different farms and because insect pests are rarely specific to one plant host and also often have strong migratory powers.

(c) *Cultivation of the ground.* By proper cultivation of the soil the insect pest is affected in two ways, *indirectly* by producing a medium for rapid growth and the production of a healthy plant, better able to withstand attack; and *directly* by disturbing the insects in their habitat and exposing them to the influence of adverse weather conditions and the attacks of their natural enemies.

That the production of a fine tilth may sometimes be disadvantageous is demonstrated by the behaviour of the cutworm. This larva is thought to have been originally a surface living creature but the cultivation of the ground has enabled it to live and progress below the top layer of soil and thus to be protected from its natural enemies. This condition may be counteracted to a certain extent by consolidating the soil surface by rolling, which tends to bring the insects to the surface. Rolling, in itself, is a useful measure against attacks by certain insects, particularly wireworms, leather-jackets and the wheat bulb fly. The effect of the rolling is to prevent easy migration through the soil from plant to plant, and, in the case of wheat bulb fly, to promote tillering of the plants. In rolling it is important to begin in time; as soon as possible after the first signs of attack are noticed. (*Journ. Min. Agric.* xxxv.)

(d) *Manuring.* By the use of manures and stimulating dressings applied at the right time the crops are helped to withstand or grow away from insect attack. A case in point is the resistance afforded to onions against attacks by *Hylemyia antiqua*, the onion fly, by liberal dressings of artificial manures. In areas where the gout fly (*Chlorops taeniopus*) is prevalent, manuring the barley is useful but must be carried out with discrimination; while small dressings of nitrogenous manures may reduce infestation, large dressings on the other hand encourage attack by retarding the growth of the ear. (See p. 235.) In some cases also, the use of farmyard manure is inadvisable as it may attract certain insect pests such as the mangold fly (*Pegomyia hyoscyami*).

(*e*) *Harvesting*. The method of harvesting the crop is sometimes important, particularly in the case of certain cereal pests, such as the Hessian fly (*Mayetiola destructor*), where the pupae may be gathered with the grain, and the wheat stem sawfly (*Cephus pygmaeus*), where the hibernating larvae remain in the stubble. The following statement in regard to harvesting and insect infestation is quoted from Wardle and Buckle (10): "It may be laid down as a general procedure that cereals should be threshed as soon as possible after harvesting and that the seed product should be stored under insect-proof conditions, and if in any way infested by pests should be disinfected as soon as possible".

(*f*) *Farm hygiene*. Clean cultivation has so often been quoted as a cure for various insect pests that it has come to be looked upon as a mere platitude. Nevertheless, though in no sense a cure for insect epidemics, farm hygiene is a valuable preventive measure. A glance through the lists of wild host plants given in this book will show how easy it is for insects to exist upon weeds till the favoured crop comes round again and thus defeat the purpose of crop rotation. Again, the cleaning out of dirty and overgrown ditches and the removal of rubbish heaps will reduce the amount of shelter for hibernating insects. In farm hygiene, then, may be included such recommendations as the removal and burning of rubbish, the destruction of unnecessary grass and the elimination, so far as practicable, of weeds and the cleaning out of ditches. In conclusion, it must be pointed out that in all these cultural methods of control, co-operation is essential; in view of the migratory powers possessed by many insects it is little use adopting these measures in isolated areas.

BIOLOGICAL CONTROL. Natural, or biological, control invokes the aid of the natural enemies, either parasite or predator, of the insect pest in question. The parasites or predators are bred artificially in large numbers and introduced into the country or district of the pest to be controlled. One of the first and most successful cases of such biological control was the almost complete eradication of the fluted scale (*Icerya purchasi*) in California by the introduction of the Australian Coccinellid beetle *Vedalia cardinalis*. Since that time the degree of control of serious insect pests attained by this method has not, perhaps, risen to expectation. There is, however, interplay of many complex factors governing the introduction and successful establishment of parasites in new districts or countries, and when these are better understood, greater success may be anticipated. Up to the present, the introduction of natural enemies of specific insects has been most successful in island areas, such as the Hawaiian Islands where many serious pests of the sugar cane have been eradicated by this means (Imms (3)). Thompson (6) gives three main reasons for this comparative failure to effect control of pests in continental areas. The first is based upon the similarity between the faunas and floras of such continental areas as Europe and North America, and it would therefore seem that the transfer of a phytophagous insect from Palearctic to Nearctic regions and vice versa should not produce any marked change in its economic status. The second reason is the phyptophagous habits of many parasites and the third is the behaviour of hyperparasites in relation to introduced primary parasites. The parasite which is most likely to survive on introduction to a new country and effect the desired control of a particular pest is one which is superior to its host in rapidity of development, adaptability to environment and powers of flight and dispersal. It should also be confined in its parasitic habit to as few host species as possible. In England, attempts at biological control have so far been confined to two insects, neither of which is of agricultural importance. The first is the woolly aphis of the apple (*Eriosoma lanigera*), and the parasite in question the Chalcid *Aphelinus mali* (Stenton (5)); the second is the greenhouse whitefly (*Trialeurodes vaporariorum*) and its parasite the Chalcid *Encarsia formosa* Gahan (Speyer (4)).

For a good general account of biological control the reader should consult Wardle and Buckle, *The Principles of Insect Control*, and Wardle, *The Problems of Applied Entomology*; while the papers of W. R. Thompson (7) (8) give an account of the principles underlying biological control.

LEGISLATIVE CONTROL. Control by legislation has two aspects; there is firstly legislation against the entry into the country of foreign pests on imported plants or other goods, and secondly legislation to control movement within a country or district of resident pests.

As regards the efforts to prevent the importation of foreign pests, there are different methods in vogue in different countries. In Great Britain the method employed is that of inspection on arrival by qualified inspectors of the Ministry of Agriculture; these are part of the Phytopathological Service described in the introductory chapter. Certain plants are regarded as especially dangerous owing to their ability to carry foreign pests; such plants are *scheduled* and can only be imported if they have received a certificate of health from the exporting country, or failing that, have been examined and released by the Ministry of Agriculture. The two pests scheduled which are of particular interest to the agriculturist are the Colorado beetle (*Leptinotarsa decemlineata* Say.) and the potato moth (*Phthorimaea operculella* Zell.). This inspection system of imported plants is reinforced by the use of the *embargo* in cases where inspection, however carried out, could not disclose the presence of some unusually dangerous or troublesome pest. The Colorado beetle is under such an embargo, no potatoes being accepted from countries in which the Colorado beetle exists. Incidentally it may be mentioned that this beetle has now appeared in the north of France and in the Departments of the Dordogne and Deux-Sèvres as well as in the south-west in the Department of Haute-Vienne and also in the Department of L'Aude. In all about fifty-three districts are now declared infected.

The second part of the legislative measures deals with the resident pest, and this has not a great application in the British Isles. The large larch sawfly (*Lygaeonematus erichsonii*) was scheduled and made compulsorily notifiable in order to obtain knowledge as to its distribution; while the nun moth (*Lymantria monacha*) was scheduled for other reasons. This latter, though present, is not a pest in the British Isles, but is a serious pest on the continent, particularly in Germany, on pines. It was therefore scheduled in the interests of those nurseries which had an export trade with America, since it was felt that this action would ensure that such nurseries at least would be free from the pest. The narcissus fly (*Merodon equestris*), already widely distributed, was scheduled as causing possible risks to the bulb trade with New Zealand. These three insects are no longer compulsorily notifiable. In 1927 the Destructive Insect and Pests Acts were amended, giving the Ministry of Agriculture powers to deal not only with insects and fungi but also with "bacteria and other vegetable or animal

organisms and any agent causative of a transmissible crop disease";
this description covers 'virus' diseases of crops which may in time
come to be scheduled. The Acts were also amended to authorise the
Ministry to pay compensation for destruction up to £2000 a year
without Treasury sanction, so that immediate action can be taken
in the case of outbreaks of dangerous foreign pests and diseases,
and the inspectors were given power to ensure the destruction of
plants attacked by foreign invaders where the owner could not or
would not do so. (Fryer(1).) These efforts towards the control of
insect pests by legislation are reinforced by the Phytopathological
Service, by means of which over sixty trained observers are available for the detection of new pests and diseases. The introduction
of legislation to deal with the warble fly pest of cattle in England
and Wales is not considered practicable at the present time. In the
first place this is due to the fact that infested animals are regularly
imported from Ireland and Canada while the work of eradication
is proceeding in this country, thus seriously prejudicing the control
measures. In the second place compulsory measures along the lines
of preventive treatment to prevent oviposition by the fly are
useless until some reliable preventive is achieved. Such legislation,
however, might be introduced were any control over importation
practicable.

REFERENCES

(1) FRYER, J. C. F. (1928). Legislation in England against diseases and pests of Plants. *Ann. App. Biol.* xv, No. 2.
(2) HOWARD, L. O. (1926). Parasite Control of insect pests. *Journ. Econ. Entom.* xix.
(3) IMMS, A. D. (1926). The biological control of insect pests and injurious plants in the Hawaiian Islands. *Ann. App. Biol.* xiii, No. 3.
(4) SPEYER, E. (1927). An important parasite of the greenhouse whitefly (*Trialeurodes vaporariorum*). *Bull. Entom. Res.* xvii.
(5) STENTON, R. (1925). Introduction of a parasite of the woolly aphis. *Journ. Min. Agric.* xxxii.
(6) THOMPSON, W. R. (1928). A contribution to the study of biological control and parasite introduction in continental areas. *Parasitology*, xx.
(7) THOMPSON, W. R. (1929). Natural control. *Parasitology*, xxi.
(8) THOMPSON, W. R. (1930). Principles of Biological Control. *Ann. App. Biol.* xvii.
(9) WARDLE, R. A. (1929). *The Problems of Applied Entomology.* Manchester University Press.
(10) WARDLE, R. A. and BUCKLE, P. (1923). *The Principles of Insect Control.* Manchester University Press.

CHAPTER III

THE EFFECT OF WEATHER CONDITIONS ON INSECT OUTBREAKS

That climatic conditions exercise a far-reaching influence on the insect pests of agricultural crops is of course well known, but knowledge as to the exact relationships existing between insects and climatic factors is at present rather scanty, and is largely confined to the influence of temperature and humidity upon certain insects in certain regions. The recent Conference of Empire Meteorologists, 1929, emphasises this important connection between weather conditions and agricultural entomology, and shows the desirability of greater co-operation between the research meteorologist and the entomologist. As knowledge of the subject increases it may be possible to forecast outbreaks of certain insect pests with some accuracy, and by statistical analysis of climatic conditions to determine the probable extent of distribution of injurious insects, particularly perhaps of those inhabiting the soil. It is already possible, by comparing the effects of a succession of cold winters for example, to gain an idea of what insect epidemics may be expected in the ensuing season. Perhaps the most outstanding instance of the correlation of insect outbreaks with meteorological factors is the case of the American 'pale western cutworm' (*Porosagrotis orthogonia* Morr.). Uvarov[12], quoting Cook[3], says: "Outbreaks of this cutworm have been found to depend on the total amount of rainfall in May, June and July of the preceding year; the critical amount of precipitation varies with the temperature and produces a definite critical soil moisture. If the moisture exceeds the critical amount, the number of cutworms decreases in the following year; if the moisture be less than the critical amount, the number of cutworms increases." The heavy rain forces this species to leave its subterranean workings and emerge on the surface where it becomes an easy prey to its parasites.

The chief climatic factors influencing the distribution and development of insects are, temperature, moisture, wind and light, and the effect so far as is known of these factors upon certain insects will be briefly dealt with in the remainder of this chapter.

TEMPERATURE. Insects, during the various phases of their life cycles, may live through a range of temperature of about 50° F., and they are also able to withstand great extremes of temperature. This reaction to extremes of temperature, however, may differ in the same insect, apparently according to the physiological condition of the insect at the time. Uvarov quotes the work of Payne(9), who has shown that the larvae of certain wood-boring beetles freeze hard at − 12·8° C., but only if the larva has been taken from its natural habitat in winter, while the same larva collected in summer freezes at − 0·77° C. This variation is thought to be due to the amount of water in an insect's body, which may vary considerably during its life; as, for example, the Colorado beetle, which according to Tower(11) loses 30 per cent. of its gross weight by loss of water before hibernation. Thus, insects seem better able to withstand low temperatures during their hibernating periods, which may be one reason why a hard winter does not necessarily mean a reduction in the numbers of certain insect pests in the following season. It appears, then, that considerable variations in temperature occurring at unusual times may find the insect in a condition unprepared for such variations, and an epidemic will be thus averted. Fox Wilson (7) quotes the case of a threatened infestation of sawfly larvae which failed to materialise through the heavy mortality of the young larvae brought about by a succession of unusually low night temperatures. Göhler(8) considers that it is *sudden* violent temperature changes which prove fatal, as native insects can withstand extreme cold provided that the transition to and from it is gradual.

It is obvious that there must be considerable interplay of the various climatic factors to produce their effect upon insect economy, and it cannot always be said that specific results are due to temperature alone; nor is the effect of temperature necessarily a direct one, it may be a secondary result arising from the reaction of temperature upon the biological agencies of control of a particular insect. For example, the aphis *Myzus persicae*, which thrives best between 55 and 80° F., may be heavily parasitised under glasshouse conditions by the Braconid *Aphidius matricariae*. This parasite, however, appears to have a smaller temperature range than the host, and although the difference is slight yet it is sufficient to allow of a more rapid increase of the aphis than would otherwise be the case. Another indirect effect of temperature upon the numbers of an insect pest consists in the destruction of the leaves of its host plant by a severe late frost, thus depriving the insect of its food supply.

A more direct effect of temperature upon insects is illustrated by the behaviour of flea beetles (*Phyllotreta* spp.); an unusually warm day occurring in early spring is sufficient to bring forth the hibernating beetles almost instantaneously in very large numbers and thus produce an epidemic. Low temperatures have as a rule a retarding, and high temperatures a stimulating, effect upon the development of insects. In the colder districts of France, the life cycle of the cockchafer (*Melolontha melolontha*) occupies four years instead of the more usual three in warmer parts, while in Russia, the moth *Phlyctaenia forficalis*, which is usually single brooded, is influenced by high temperatures to produce a partial second brood.

MOISTURE. Bachmetjev[1] makes the following statements in regard to the effect of moisture upon insects: (1) that there is an optimum degree of moisture for insect development, (2) that this optimum is not the same for different species, (3) that the moisture which may hasten the development of one species, may retard the development of another.

The effects of moisture upon insect pests of English agricultural crops can be illustrated by various examples. The diamond-back moth (*Plutella maculipennis*) is very susceptible to moisture and only becomes abundant in hot dry seasons. Its development is retarded by wet cold weather and a heavy rainstorm is often sufficient to prevent an attack developing. Davies[4] has shown that a very high percentage of moisture is necessary for the development of certain species of Collembola (*Bourletiella hortensis*). These insects may be found in large numbers attacking mangolds in the early hours of the morning, but none occur during the heat of the day. Control measures against *B. hortensis*, therefore, must be applied in the early morning or during wet weather. Climatic conditions, particularly moisture, influence the numbers of cut-worms (*Noctuidae*) to a large extent, and a factor governing the occurrence in epidemic form appears to be the rainfall in late spring, while heavy rains in July seem to be instrumental in checking an attack. The wireworm, on the other hand, is peculiarly prone in all its stages to desiccation, as is also the leatherjacket (*Tipula*), and, as regards the latter, farmers consider that a hot and dry August and September reduce the numbers of overwintering larvae, while a wet autumn is favourable to their development. It is likely, also, that the effect of rain on insects is partly mechanical, due to dislodgment from the plant host and death by drowning or injury. In 1925 a serious early attack of the cabbage aphis (*Brevicoryne brassicae*) was checked by heavy rain in August. The indirect effect of moisture upon insect epidemics by its influence, beneficial or otherwise, upon the growth of the host plant is also very great.

The combined effect of temperature and moisture upon an important agricultural pest, the mangold fly (*Pegomyia hyoscyami*), is illustrated by the work of Blunck, Bremer and Kauffman [2] in Germany, who find that an analysis of the fluctuations of infestation indicates that the most important restricting factors are summer temperature in its indirect effect on the development of the parasites and summer rainfall when it is definitely in excess of the normal. Heavy rain washes away the eggs and kills the pupae. In Central and Northern Europe, low summer temperatures prevent the coincident occurrence of parasites and host. Epidemics are likely to occur when the temperature is such as to favour development of the fly and at the same time is cool enough to be unfavourable for an equally rapid development of the parasites.

WIND. It is probable that wind plays a large part in the dispersal of insect pests, being either favourable to the spread of some or preventing the spread of others in certain directions. The Hessian fly (*Mayetiola destructor*), among others, seems to be largely dependent upon winds for its dispersal. It has been shown also that aphides are sometimes carried enormous distances by means of the wind. Experiment has proved that winged aphides are easily carried to Heligoland from the mainland, a distance of nearly 50 miles, and Uvarov [12] quotes the case of the appearance of winged spruce aphides on the snow at Spitzbergen where they could have been brought only by wind from the Kola Peninsula, a distance of over 800 miles in a straight line (Elton [6]).

This transportation of aphides by the wind has a practical bearing on the question of potato virus diseases, when attempts are being made to grow stocks of virus-free potatoes in areas far from other potatoes and away from virus-bearing aphis vectors. (See Chapter xiv.)

LIGHT. Little is known of the precise effects of light upon insect activity. De Gryse(5) describes how attacks by the white pine weevil in Canada may be forestalled in reafforestation projects by planting the pines under cover of fast-growing broad-leaved species, thus preventing egg-laying by the sun-loving adults. Some facts have also been obtained as to the influence of light upon the production of the sexual forms in some species of aphides. Sunlight appears to influence the egg-laying activities of certain parasitic flies. Of the two species of warble flies for example, *Hypoderma bovis* and *H. lineatum*, the former seems only to oviposit in sunny weather while the latter will lay eggs freely at temperatures of 55° F. The sheep nostril fly, *Cephalomyia ovis*, and the cattle fly, *Hippobosca equina*, also seem only to fly and reproduce in bright sunshine. Light appears to have a marked influence on the development of the larvae of *Stomoxys calcitrans* (p. 210).

In conclusion, the following is quoted from the work of W. R. Thompson(10): "Insect outbreaks, considered in general, have no uniform and common cause in any environmental condition or combination of environmental factors. The only absolutely general statement that can be made in regard to them is that they arise because the environment has for a time approached the optimum for the species concerned. Since the environment in which the organism is in control may be of many different types, the approximation to the optimum may be from many different directions and may differ in degree. Thus in an environment which is too dry, a little more moisture may give rise to an outbreak; in one which is too cold, a little more heat will have the same effect. On the other hand, all environments in which outbreaks occur will not necessarily be similar. Thus, supposing that a certain species is kept in control in one region by the combined action of several parasites and some form of agricultural control, and in another by an absence of moisture together with an unfavourable distribution of the food plant. The disappearance of one of the parasites from the first environment and an increase of moisture in the second, might lead to simultaneous outbreaks in both cases. Outbreaks of species are simply due to the fact that conditions momentarily correspond or approximate to the ecological optimum; control means simply a departure in one or many directions from optimum conditions to the point where the species is just able to maintain itself".

18   REFERENCES

REFERENCES

(1) BACHMETJEV, P. (1907). *Experimentelle entomologische Studien vom physikalisch-chemischen Standpunkt.* Ausg. 2.

(2) BLUNCK, H., BREMER, H. and KAUFMAN, O. (1929). Untersuchungen zur Lebensgeschichte und Bekämpfung der Rübenfliege (*P. hyoscyami* Pz.). *Arb. biol. Reichsanst. Land. u. Forstw.* XVII, No. 2.

(3) COOK, W. C. (1927). Some effects of alternating temperatures on the metabolism of cutworm larvae. *Journ. Econ. Entom.* XX.

(4) DAVIES, W. M. (1928). The effect of humidity on Collembola. *Brit. Journ. Exper. Biology,* VI, No. 1.

(5) DE GRYSE, J. J. (1929). The relations of entomology to meteorology. *Conference of Empire Meteorologists, Agric. Section.* II. *Papers and Discussions.* H.M. Stationery Office.

(6) ELTON, C. S. (1925). The dispersal of insects to Spitzbergen. *Trans. Entom. Soc. London.*

(7) FOX WILSON, G. (1929). *Conference of Empire Meteorologists, Agric. Section.* II. *Papers and Discussions,* p. 161. H.M. Stationery Office.

(8) GÖHLER, H. (1929). Ist dieser Winter ein Schadlingsvernichter? *Die Kranke Pflanze,* VI, No. 3. Dresden.

(9) PAYNE, N. (1927). Freezing and survival of insects at low temperatures. *Journ. Morph.* XLIII.

(10) THOMPSON, W. R. (1929). Natural Control. *Parasitology,* XXI.

(11) TOWER, W. L. (1917). Inheritable modification of the water relation in the hibernation of *Leptinotarsa decemlineata. Woods Hole Bulletin.*

(12) UVAROV, B. P. (1929). Weather and climate in their relation to insects. *Conference of Empire Meteorologists, Agric. Section.* II. *Papers and Discussions.* H.M. Stationery Office.

CLASSIFICATION. COLLEMBOLA; ANOPLURA; THYSANOPTERA

The classification adopted in the present volume is that given by Imms in his *General Textbook of Entomology* and the orders of agricultural importance in the British Isles are marked in heavy type.

Sub-class 1. APTERYGOTA.
  Order 1. Thysanura. Bristle-tails.
      2. Protura.
      3. **Collembola.** Spring-tails.

Sub-class 2. PTERYGOTA.
  Division 1. EXOPTERYGOTA.
  Order 4. Orthoptera. Cockroaches, Grasshoppers, etc.
      5. Dermaptera. Earwigs.
      6. Plecoptera. Stone Flies.
      7. Isoptera. Termites.
      8. Embioptera.
      9. Psocoptera. Book-lice.
     10. **Anoplura.** Biting and Sucking Lice.
     11. Ephemeroptera. May Flies.
     12. Odonata. Dragon Flies.

Order 13. Thysanoptera. Thrips.
     14. **Hemiptera.** Plant bugs, aphides, etc.

Division 2. ENDOPTERYGOTA.
  Order 15. Neuroptera. Alder Flies, Lacewings, etc.
     16. Mecoptera. Scorpion Flies.
     17. Trichoptera. Caddis Flies.
     18. **Lepidoptera.** Butterflies and Moths.
     19. **Coleoptera.** Beetles.
     20. Strepsiptera. 'Stylops.'
     21. **Hymenoptera.** Sawflies, Ants, Bees, Wasps.
     22. **Diptera.** Two-winged or True Flies.
     23. Aphaniptera. Fleas.

Sub-Class 1. APTERYGOTA

In this division are placed three orders of wingless insects, only one of which, the Collembola, contains a genus of agricultural importance.

Order 3. **Collembola.** Spring-tails

Small insects elongate or globular in form, occurring in damp soil, under bark of trees, etc., often specifically social, characterised by the possession of a ventral tube on the first, and a spring attached to the fourth, abdominal segment. Eyes sometimes absent, antennae usually elbowed, only six abdominal segments. The spring, which gives this group its popular name, is carried deflexed, pointing forwards under the abdomen. The insect leaps by means of this apparatus, which is probably not directive in its action.

There are two species of Collembola of some agricultural importance, i.e. *Bourletiella* (*Sminthurus*) *hortensis* Fitch and *Sminthurus viridis* Lubb. Often associated with these are other species which are mentioned shortly.

Bourletiella (*Sminthurus*) *hortensis* Fitch

DESCRIPTION. Body globular in shape, colour blackish to dull green above, usually marked with white spots. Eye spots large and black sometimes bordered with white or yellow. At the base of the

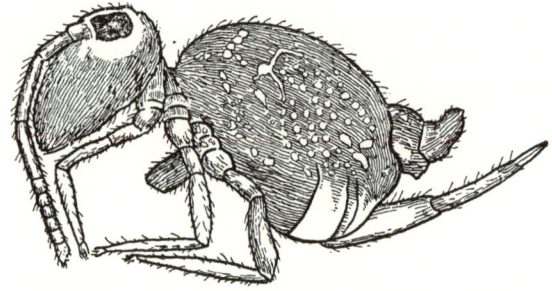

Fig. 2. *Bourletiella hortensis* Fitch. Male. × 55. (After Folsom.)

spring there is often a white or yellowish area. Antennae ferruginous, about one and a half times as long as the head which is reddish brown above. Legs dull blue, trochanters bluish or brownish, femora with white spots. Ventral tube blue, from which may be protruded a pair of long tubules, each tuberculate on the distal half. Dense short curving white setae on head and abdomen becoming longer on the anal segment. Length 1·2–1·5 mm. (Abbreviated from Folsom.) (Fig. 2.)

CULTIVATED HOST PLANTS. Chiefly mangolds, beet seedlings, occasionally sugar-beet, potatoes, soy beans, radishes and peas. The mangold variety Yellow Globe is particularly susceptible,

Golden Tankard much less so (Davies), also reported as sporadically attacking many other plant hosts.

WILD HOST PLANTS. Groundsel (*Senecio vulgaris*), goosefoot
(*Chenopodium* spp.) and redshank (*Polygonum persicaria*).

SYMPTOMS OF ATTACK. As it is largely the damage caused to young
mangolds which renders this insect of agricultural importance, it is
in relation to that plant that the symptoms of attack are considered.
Damage may be done to both leaf and stem, in the former round
holes are eaten somewhat reminiscent of flea beetle attack (see
p. 119). As regards the latter the process of feeding on the stem is
thought to give rise to a condition in young mangold plants known
as 'blackleg' or 'strangle'. In this condition the stem is eaten
round, at, or just above, the ground level, giving it a constricted
appearance, the crown above this constriction and the root in the
soil below it, are undamaged. Usually a large number of the
insects concentrate, feeding on one spot, thereby keeping the wound
open and causing considerable 'bleeding'. This should not be confused with somewhat similar damage caused by *Atomaria linearis*
(see p. 94) in which the constriction lies below the soil level. The
amount of damage done varies, but may reach 20 to 30 per cent. of
the crop. Young plants affected with this characteristic constriction
are very liable to be broken off by the wind. Damage becomes
noticeable at the end of June and beginning of July, and the insects
have mostly left the crop by the middle of August. It should
perhaps be mentioned that the precise connection of spring-tails
with 'strangle' is not yet determined.

DISTRIBUTION. Widely distributed throughout the British Isles,
attacks occurred in Nottinghamshire, Yorkshire, Bedfordshire,
Hertfordshire and South Wales in 1926 and in Yorkshire and
Shropshire in 1927.

CONTROL. Davies(2) has shown that large numbers of Collembola may be caught on tarred sacks stretched across a light wooden
frame mounted on wheels and pushed by the operator along the
rows. This apparatus is best used in dull, damp weather which is
favourable to spring-tails. If an insecticide is used, a 5 per cent.
nicotine sulphate dust sometimes gives good results. A 1 per cent.
green tar oil dust proved an active repellent and fine air-slaked lime
a good contact insecticide. Earthing up the young mangolds to
protect the stem from attack has been suggested as a possible

preventive. Any control measures against *B. hortensis* must be applied in the evening or early morning when the humidity is high; they are of no use applied during the heat of the day.

Sminthurus viridis Lubbock

DESCRIPTION AND LIFE HISTORY. A small yellow globular species which occurs commonly on grass land and clover, and may do considerable damage to the latter crop. The eggs are laid in batches of 30–40, usually on damp decaying leaves or other moist surfaces at the base of the plant. These hatch in 20–25 days, the insect being mature in 8–10 weeks.

HOST PLANTS. The following grasses, etc., are favourite hosts: wild white clover (*Trifolium repens*), late flowering clover (*T. pratense*), rye grasses (*Lolium* spp.), cocksfoot (*Dactylis glomerata*), and timothy grass (*Phleum pratense*). *S. viridis* also occurs on root crops and cereals.

CONTROL. Heavy rolling and chain harrowing are recommended, the attachment of tarred sacks to the front of the roller should render the former doubly effective.

Associated with the foregoing two species are the following:

Bourletiella lutea Lubbock

A small yellow globular species, almost universally present on potato, usually on the under sides of the leaves.

Isotoma viridus Bourlet

Isotomerus palustris Muller

Small elongate species mostly brown or slate coloured, often found in the soil surrounding mangold plants. None of these last three species is of great agricultural importance.

Sub-Class 2. PTERYGOTA

Division 1. EXOPTERYGOTA

Order 10. **Anoplura**. Biting and Sucking Lice

Small insects, flattened and apterous, ectoparasitic on birds and mammals, feet adapted for clinging; abdomen without cerci; no metamorphosis.

Sub-order 2. *Siphunculata* (*Anopleura* of most writers).
Sucking Lice

One species of agricultural importance is dealt with shortly; this
is the hog louse, *Haematopinus suis* Linn.

DESCRIPTION. *Adult. H. suis* is the largest of the lice; the male and
female differ in size, in the shape of the abdomen and in the structure of the two posterior abdominal segments. The abdomen of the
male is shorter than that of the female, and there is a transverse row
of hairs on each abdominal segment. There is a strongly chitinised
plate of characteristic shape extending from segment 7 to segment
9 on the ventral surface of the abdomen. The longer abdomen in
the female gives the insect a more slender appearance, there are
fewer hairs, less regularly arranged. The ninth segment has a deep
indentation together with lateral chitinised processes which are
used for clasping the host's bristles; on the dorsal surface of the
segment is a strongly chitinised plate. (Florence (5).) Length, male
3·9 mm., female 4·6 mm.

Egg. White, changing to amber before hatching; symmetrical,
tapering posteriorly, bluntly rounded at the anterior end, surface
punctate. Length 1·5–1·75 mm.

Early stages. According to Florence, the newly hatched larva
has 5-segmented antennae and a 9-segmented abdomen. There
are three instars, maturity being indicated at the third moult by the
appearance on the thorax of a sternal chitinous plate.

LIFE HISTORY. The hog appears to be the only host of this louse,
although man does occasionally act as a temporary host. On the
hog the lice are found mostly on the folds of skin on the neck and
jowl, round about the ears and the inside of the legs and the
flanks. The eggs are deposited one at a time on the bristles of the
hog and attached with a clear cement, the incubation period being
12–20 days, according to the temperature. The whole life cycle
appears to occupy 29–33 days.

CONTROL. Ill-fed animals and those in dirty pens are most
usually affected. Hadwen recommends raw linseed oil rubbed on
with a brush; kerosene emulsion is also effective. Pigs should not
be left out in the sun after treatment as the skin is liable to injury
by scalding.

Order 13. Thysanoptera. Thrips

Very small elongate insects, wings when present two pairs, long narrow and fringed with little venation. Mouth parts suctorial, asymmetrical, the needle-like mandibles are enclosed by the conical beak. Ten abdominal segments, antennae 6–9-jointed, tarsi 2-jointed and bladder-like at the tip. One or two species are of agricultural importance.

Frankliniella robusta Uzel. Pea Thrips; Bean Thrips; 'Black Fly'

This species belongs to the sub-order Terebrantia and may be recognised by the terminal segments of the abdomen which are constricted to a point and not tubular as in another sub-order, Tubulifera. The female possesses a saw-like ovipositor.

DESCRIPTION. The following descriptions of the adult male and female pea thrips are given according to Williams (14).

Female. Length 1·85 mm. Body, legs and antennae dark brown, except for the third and base of the fourth antennal segments, tarsi and fore tibiae which are paler; fore wings heavily tinged with brown, lighter at the base, hind wings almost transparent; head broader than long, cheeks almost parallel; eyes dark, not protruding, ocelli present forming an equilateral triangle; two small setae in front of the anterior ocellus and one on each side near the margin of the eye. The two long ocellar spines are between the posterior ocelli, a long spine behind each eye and two pairs of short ones in a row between these; a few short forwardly directed spines on the cheeks; faint striations near the margin of the head. Mouth cone short and rounded, reaching about three fifths across the prosternum. Maxillary palps 3-segmented, the second segment shorter and the third longer, than the first. Labial palps 2-segmented, the basal joint very short, antennae more than twice as long as the head; first segment short, broad and barrel-shaped; second, broad and truncate at apex; third, the longest, with short pedicel and dorsal forked trichome; fourth, spindle-shaped, slightly broader at the apex than at the base, with ventral forked trichome; fifth, narrowing slightly in the distal third and then truncate; sixth, spindle-shaped; seventh, shorter and broader than the eighth.

Prothorax wider than long and longer and wider than the head. Two long spines at each hind angle and one at each front angle, also two not quite so long on the front margin and four or five pairs of short ones on the hind margin.

Legs normal, all tarsi yellow, fore tibiae paler than femora; two rows of stout spines on the hind tibiae.

Wings reaching to the ninth abdominal segment; fore wings clouded with brown except at the base. Hind wings clear, the single vein indistinct but distinguishable to near the tip of the wing.

Abdomen, the ventral pleurites pectinate posteriorly; there is a very short projection on each side of the eighth abdominal segment (corresponding position to the much larger ones in the male). A row of short pointed tooth-like projections on the hind margin of the eighth tergite. The tenth segment longer than the ninth and split dorsally for about three quarters of its length from behind.

Male. About one sixth smaller than the female. All the antennal segments, especially the first two, much paler than in the female. On each side of the eighth segment is a short process curving backwards and upwards and ending in a blunt point.

Egg. Very small and soft, white in colour, bean or kidney-shaped.

LIFE HISTORY. The adult thrips begin to appear about the middle of May. They may be found in the terminal leaf clusters and unopened flower buds of peas, and continue to occur until about the end of July or beginning of August, the period of greatest infestation being in the middle of June. The egg is inserted into the tissue of the plant in a slit made by the ovipositor. The majority of the eggs are laid in the soft and succulent sheath round the stigma or young pod formed by the united stamens, more rarely in the flower petals. The egg stage lasts 7–10 days. There are two stages in the development of the larvae, occupying about 17 days, the second stage larva is bright orange with the last two abdominal segments dark brown. The larvae become full fed during May and June when they leave the host plant and descend into the soil, the winter is passed in the larval condition in cracks and crevices in the soil. The following May and June the larva passes through two short stages occupying only about two days, firstly the propupa in which the wing rudiments appear, this rapidly passes into the second or pupal stage in which the head, thorax and abdomen are bright orange-yellow with antennae, legs and wing cases almost colourless, the last two abdominal segments are pale and not dark brown as in the larva. The adult thrips appears again in May, there is thus only one brood per annum.

CULTIVATED HOST PLANTS. Most commonly on the edible pea (*Pisum sativum*) but broad beans (*Vicia faba*) are also attacked. Both field and garden crops may be affected.

WILD HOST PLANTS. Knapweed (*Centaurea nigra*).

SYMPTOMS OF ATTACK AND INJURY TO HOST PLANT. The flowers, pods and terminal shoots suffer most from attack by the pea thrips. The pods are undersized and sickly looking, presenting the characteristic silvery appearance caused by the punctures of thrips. (Wardle and Simpson (13).) (Fig. 3.) Occasionally the flower itself is so injured as to inhibit further development.

OTHER SPECIES OF THRIPS OCCURRING ON LEGUMINOUS PLANTS. Three other species of thrips occasionally occur in company with *F. robusta*. These are:

Aelothrips fasciatus. Recognised by larger size and banded black and white wings.

Frankliniella intonsa. Common on scarlet runners, distinguished from *F. robusta* by the absence of the lateral abdominal processes.

Thrips tabaci. Antennae 7-segmented, colour greyish to brown. Male lighter.

DISTRIBUTION. Much loss was caused by *F. robusta* in south Lincolnshire in 1925 and it was harmful to mid-season crops in the south-eastern provinces. In 1926 it was less harmful and only abnormally prevalent in Somerset and Devon. In 1927, except in the northern and south-western parts of the country, the pest was extremely abundant and destructive, in some districts causing complete loss of the crop.

CONTROL. Spraying is only of use when the larvae are exposed on large pods, in such cases a D.D.T. dust or emulsion might be effective. Fields which have been affected during the summer should be ploughed deeply and receive a dressing of lime or soil fumigant in the winter to kill the hibernating larvae. Owing to the habit of the female of ovipositing in the stamen sheath of the flowers early sown peas are more likely to escape infection. Wet weather causes heavy mortality upon thrips generally though less so on this species, owing to the protection afforded by the flowers. light soils are favourable to this thrips.

NATURAL ENEMIES. The Chalcid *Thripoctenus brui* is recorded from *F. robusta* in France.

Fig. 3. Pea pods damaged by thrips, *Frankliniella robusta* Uzel. (Reproduced by permission of the Ministry of Agriculture.)

Limothrips cerealium Hal. Corn Thrips

DESCRIPTION. *Adult*. Smooth, shining, pitchy black, flattened; female winged, male apterous. Head oval with three ocelli triangularly placed. Antennae 9-jointed, bristly, mouth parts form a short black beak. Thorax almost square with four dots, two on each side. Abdomen 9-segmented, long narrow and smooth, last segment with two lateral spines in the male, tapering in the female with a flattened ovipositor. The female has two pairs of pale ciliated wings. Legs, short, straw-coloured. Length, 2 mm. (Curtis[7].) (Sharga, (1933), *Ann. Appl. Biol.* xx, 308–326.)

Larva and pupa. Similar to adult, but smaller and yellow in colour, pupa paler, eyes reddish.

LIFE HISTORY. The winter is probably passed as the adult female, or possibly in the pupal stage in the soil, thick grass, stubble, etc. The winged females appear on grasses about April and May, later they oviposit on oats, barley and other cereals, the eggs being deposited in the tissue of the outer glumes. These hatch in a few days and the insects feed as larvae and adults within the ear sucking the juices and causing shrivelling and discoloration of the grain. One or more seeds in an ear of corn may be affected by 'blindness', the glumes appearing whitish and paler than the rest of the ear. The thrips become adult about the middle of August and remain in the ear till the grain becomes too hard for them to pierce. According to Morison[9] tall grasses around margins of fields and on waste land probably provide breeding places.

CONTROL. Control measures are not applicable once the thrips are in the grain. Theobald[11] recommends the following preventive measures to destroy the hibernating insects: harrowing and burning stubble in early spring and the destruction so far as possible of wild grasses about the fields in autumn or early spring. Early sowing of spring oats in order that the plants may be well advanced before a possible attack begins is also recommended.

REFERENCES

COLLEMBOLA

(1) BRITTAIN, W. H. (1924). The garden springtail (*Sminthurus hortensis* Fitch) as a crop pest. *Proc. Acadian Entom. Soc.* for 1923, No. 9.
(2) DAVIES, W. M. (1926–7). Collembola injuring leaves of mangold seedlings. *Bull. Entom. Res.* XVII.
(3) DAVIES, W. M. (1925). Investigations of springtails attacking mangolds. *Journ. Min. Agric.* XXXII.
(4) DAVIES, W. M. (1928). On the economic status and bionomics of *Sminthurus viridis* Lubb. *Bull. Entom. Res.* XVIII.

ANOPLURA

(5) FLORENCE, L. (1921). The hog louse, *Haematopinus suis* Linn. Its biology, anatomy and histology. *Agric. Exp. Station, Cornell Univ. Mem.* No. 51, Ithaca, N.Y.
(6) WATTS, H. R. (1918). The hog louse. *Agric. Exp. Sta. Univ. Tennessee, Bull.* No. 1.

THYSANOPTERA

(7) CURTIS, J. (1860). *Farm Insects.*
(8) MILLARD, W. A. and BURGESS, R. (1927). *Blindness in barley.* Univ. Leeds and Yorks. Council Agric. Educ. No. 151.
(9) MORISON, G. D. (1928). Observations and records of some Thysanoptera from Great Britain. *Entom. Mon. Mag.* LXIV.
(10) Pea and bean thrips or black fly (1914). (Anon.) *Journ. Board Agric.* XXI.
(11) THEOBALD, F. V. (1922). Thrips in corn. *Journ. Kent Farmers' Union,* XII, 2.
(12) VUILLET, A. (1914). La thripsose de pois. *Revue Scientifique,* Paris.
(13) WARDLE, R. A. and SIMPSON, R. (1927). The biology of Thysanoptera with reference to the cotton plant. *Ann. App. Biol.* XIV.
(14) WILLIAMS, C. B. (1914). The pea thrips. *Ann. App. Biol.* I.

CHAPTER V

Order 14. Hemiptera (Rhynchota). Plant Bugs, Aphides, Coccids, etc.

The chief character common to all the members of this order is the possession of sucking mouth parts, two pairs of needle-like stylets enclosed in a groove of the modified labium. Two pairs of wings are usually present, the different modifications of which divide the group into two-well marked divisions or sub-orders—the *Heteroptera* and *Homoptera*; metamorphosis incomplete and gradual.

The order *Hemiptera* contains many well-known and important plant pests—Capsid bugs, leaf hoppers, aphides, white flies, scale insects, to mention a few. It is of direct importance to man on account of the immense amount of damage to crops arising from the activities of its members. This damage is two-fold—directly on account of the loss of sap due to the continuous feeding, and indirectly on account of the fact that many of the plant-sucking insects act as vectors of different plant diseases, inoculating the plant with the pathogene in the act of feeding. Of paramount importance among the latter is the group of plant maladies known as 'virus diseases' which are dealt with in Chapter XIV.

Fig. 4. Shoots of potato plant damaged by *Lygus pabulinus* Linn., the common green capsid bug. About half natural size.

In addition to the two-fold damage above mentioned, a third form of injury arises from toxins contained in the saliva of certain Capsid bugs which, when injected into the plant during the process of feeding, causes severe injury to the plant host. (Smith(11).) (Fig. 4.)

Sub-order 1. *Heteroptera*

Fore wings modified into wing covers, proximally horny with membranous extremities which overlap on the back. Metamorphosis incomplete, the young resembling the adult except for the absence of wings which arise from external buds as growth proceeds.

Family *Capsidae*. Capsid Bugs

The largest family of the *Heteroptera*; mostly green or brown insects which run with great rapidity over the plant when disturbed. Characterised by the structure of the wing covers which possess a cuneus, the small triangular area at the end of the wing covers, but no embolium, which is a narrow sclerite extending along the anterior margin of the wing cover or hemelytra.

Calocoris norvegicus Gmel. (*bipunctatus* Fab.) Potato Bug

DESCRIPTION. Green, clothed with rather thick black hairs; pronotum generally with two small black spots on the disc; elytra in the male sometimes more or less reddish brown between the nerves; membranes dusky, nervures pale; femora more or less freckled with brown. Basal joint of the antennae scarcely thickened, third and fourth slenderer than the second, together about equalling it in length; pronotum shining, much raised posteriorly, with a narrow raised collar, sides nearly straight, surface slightly wrinkled transversely; elytra with the sides slightly rounded; posterior femora with the sides slightly swollen. Length 7 mm. (Saunders[10].) The young stages are wingless and almost wholly green.

Egg. Yellowish in colour, cylindrical, curved, with the collar rising higher on the convex side than on the concave, its margin being crenulated, rounded posteriorly, and obliquely truncate anteriorly. Length $1\frac{3}{4}$ mm. (Butler[2].)

LIFE HISTORY. The complete life history of *C. norvegicus* in this country does not appear to be known, but the following facts are offered, based upon observations made by the writer. The early stages of this Capsid are to be found in May upon young potato plants, but usually only upon those rows bordering the hedges. Probably the overwintering eggs are inserted in the woody stems of hedgerow plants, whence the young Capsids migrate to potatoes in the following spring. In Ireland, Rhynehart[8] has observed the bugs ovipositing from August onwards in the stems of corn marigold (*Chrysanthemum segetum*); ragwort (*Senecio jacobaea*); thistle (*Carduus* spp.); redshank (*Persicaria* sp.). *C. norvegicus*, like some other allied Capsid bugs, appears to have developed in a small degree the migratory habit so characteristic of the life history of aphides. The summer is spent upon herbaceous plants, particularly potatoes and nettles, upon which oviposition may take place. It is not certainly known whether there are two broods in the year, but this is probably the case as the writer has observed *C. norvegicus* in captivity ovipositing in the stems of potato plants. The strong curved ovipositor is inserted up to the hilt in the stem and the egg so placed is invisible from the exterior. (Smith (12).) The second brood would thus oviposit in woody plants, the winter being passed in the egg stage. Massee and Steer (*Entom. Mon. Mag.* LXIV, 1928, p. 207) relate having seen this Capsid ovipositing in posts of chestnut wood.

HOST PLANTS. Though less omnivorous than the common green Capsid bug, *C. norvegicus* is found upon a very large number of plants, both wild and cultivated. From the agricultural point of view, the potato and sometimes the hop are the only crops likely to suffer from the attacks of this insect; though it has been recorded as attacking red clover in Wales and in Ireland it is frequently a pest of flax. The chief wild host plant is the common stinging nettle (*Urtica dioica*), and the Capsid can be found in great numbers upon this plant throughout the summer.

Injury to host plant and symptoms of attack. There is no doubt that the injury caused by the toxins contained in the saliva of *C. norvegicus* is the most important factor in the damage done by this insect. Attack can readily be recognised; at an early stage of infestation the potato plant shows, particularly on the young leaves and shoots, numbers of small red spots, each spot marking a puncture. These spots increase in size to a certain extent and turn brown, while holes may appear later in the leaves, the shoots are killed and the whole plant presents a distorted aspect, a condition known in some parts of the country as 'stung'. It should be noted that this damage is due entirely to the toxic nature of the saliva injected by the insect while feeding, and has no connection with virus diseases; any possible relationship of this Capsid with the spread of potato viruses is discussed in Chapter XIV. The Capsid is equally injurious in all its stages.

DISTRIBUTION. Very common all over Great Britain, being recorded from thirty-three counties; it is also well known as a pest in Ireland.

CONTROL. Efforts at the control of Capsid bugs in this country have been mainly directed against the bugs attacking fruit trees. From a survey of the literature dealing with Capsid bug control, the most efficacious spray fluid seems to be either a soap and nicotine wash or a nicotine-paraffin emulsion. To be effective, these solutions must wet the bug and herein lies their weak point. About the end of June the bug becomes winged and at the slightest alarm flies away and so avoids contact with the spray. The spray must therefore be applied soon after the first appearance of the insects while they are still in the wingless stage.

NATURAL ENEMIES. Little is known of the natural enemies of *C. norvegicus*. On several occasions the writer has found nymphs of this Capsid with a Hymenopterous larva, probably a Braconid, within them.

Lygus pabulinus Linn. The Common Green Capsid Bug

DESCRIPTION. *Adult.* Smaller than the foregoing species and almost entirely green. Bright green, shining, elongate, oval, punctured, clothed with dusky hairs: second segment of antennae green at base, testaceous in middle, black at tip, third and fourth segments black; pronotum with a narrow raised collar, sides nearly straight, base rounded, anterior callosities moderately developed; scutellum with transverse wrinkles; elytra elongate, sides slightly curved; membrane iridescent with a longitudinal cloud below the apex of the cells, veins pale green; femora elongate; tibiae with moderately developed spines. (Petherbridge and Thorpe[7].) Length 6 mm. (Fig. 5.)

Egg. Creamy white in colour, surface smooth and curved along its length with the distal end bluntly rounded and thick; from here it expands markedly on one surface and then gradually narrows to a neck-like region; the other surface is only slightly concave and narrows only in the immediate proximity of the neck. From the neck, the egg expands slightly to form a cap; this latter is barely darker than the body of the egg, being pale yellowish green in colour. It has a series of very fine longitudinal striations all round which appear to radiate from the disc of the egg. The egg is closed by a brown coloured disc. Length 1·26 mm. (Austin[1].)

LIFE HISTORY. From a survey of the habits of this bug, it is possible to pick out, in its peregrinations among a diversity of plant hosts, a well-defined cycle of movements. In April and May the young bugs hatch from the eggs, overwintering in woody plants (currants, gooseberries or apples) and migrate to herbaceous plants of many kinds, the most important being the potato. In addition to this *L. pabulinus* is frequently found upon strawberries, and it attacks many weeds, among which the bindweed (*Convolvulus arvensis*) and the stinging nettle are the most commonly affected. Probably on account of this omnivorous habit, the Capsid can hardly be considered as a very serious potato pest, though its developing taste for a fruitarian diet may soon demand more

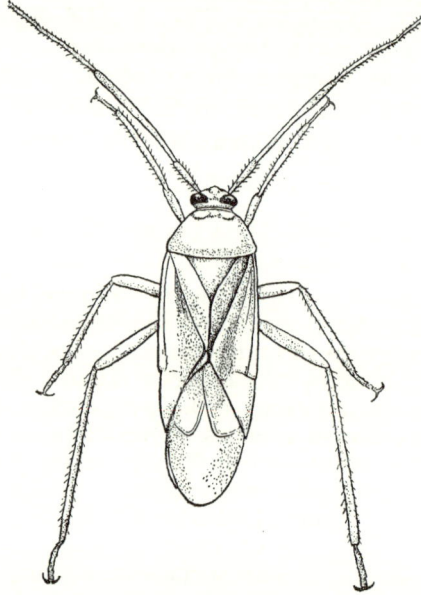

Fig. 5. *Lygus pabulinus* Linn. The common green capsid bug. × 9. (After Rostrup and Thomsen.)

attention from the orchardist. The eggs of the summer generation are deposited in the tissues of herbaceous plants, during June and July. When mature, the bugs arising from these eggs migrate once more to the woody plants, in the stems of which the overwintering eggs are deposited. There are therefore two broods of this Capsid bug in the year.

CULTIVATED HOST PLANTS. A few only of the more important cultivated plant hosts can be given here; excluding fruit these are: potato, beans (dwarf, broad and runners), sugar-beet, artichokes (*Helianthus*) and many cultivated garden plants.

WILD HOST PLANTS. The lesser bindweed (*Convolvulus arvensis*), the larger bindweed (*C. sepium*), the curled dock (*Rumex crispus*), bittersweet (*Solanum dulcamara*), black nightshade (*S. nigrum*), common sow thistle (*Sonchus oleraceus*), stinging nettle (*Urtica dioica*). The effect on the host plant and the measures for control are the same as for *C. norvegicus*.

Lygus pratensis Linn. The Tarnished Plant Bug

DESCRIPTION. Very variable in colour, being generally brown to reddish brown or yellow. Head yellowish brown, usually marked with three longitudinal lines brownish, black or reddish in colour, very variable in distinctness. Prothorax bronzy brown, usually with four more or less distinct blackish spots in a row, one-third the distance from the front margin, sometimes so arranged as to give the prothorax the appearance of having four longitudinal dark stripes; the posterior angles of the prothorax are marked with a dark brown, black or red spot. Scutellum varies from brown to black and is usually marked with a heart-shaped or Y-shaped spot on the posterior half. Wings bronzy brown, mottled with yellowish brown and reddish. Antennae dark brown, first segment and second segment except the tip usually lighter. Legs light yellowish brown to dark reddish; under surface of insect dark in the centre with a lighter stripe on each side. There is a marginal band of brown in which is a sub-marginal row of yellowish spots, one to each segment. As a rule the male is smaller and darker than the female. The ovipositor, when at rest, lies nearly concealed in a groove on the under surface of the abdomen. (Crosby and Leonard (3).) Length 5–6 mm.

Egg. Varies slightly in length and diameter, average measurements being 1 mm. long and 0·24 mm. diameter at the basal end. General outline flask-shaped and slightly curved, base somewhat enlarged and rounded; the apex is narrowed and rimmed, the margin of the rim being thickened and darker in colour; at first transparent later becoming pale yellow. (Fox Wilson(4).)

LIFE HISTORY. The habits of *L. pratensis* in this country are not accurately known. From observations made by the writer, it would appear that the winter can be passed in two or three ways: in the heated glasshouse where the insect is often a pest of chrysanthemums, as the hibernating adult out of doors, or possibly as an overwintering egg. As regards the hibernating adults the writer has found them in December under leaves and at the bottom of grass roots. When brought into a warm glasshouse they revived and commenced to feed. The adults of this species also appear able to remain active out of doors until very late in the season. During development the insect moults five times, the wings being acquired at the fifth moult. It is very plentiful throughout the summer, often occurring in company with *L. pabulinus* and *C. norvegicus*.

HOST PLANTS. Like the two foregoing species, *L. pratensis* is almost omnivorous. It attacks a variety of crops, fruit trees and ornamental plants. It has been recorded as a pest of potatoes, beans, beet, celery and occasionally Brassicas. It occurs commonly on the stinging nettle and may be found in large numbers in September on the black nightshade (*Solanum nigrum*).

SYMPTOMS OF ATTACK AND INJURY TO HOST PLANT. This insect produces very similar symptoms to those of the other two Capsids, the saliva being equally toxic to the majority of plants. It is worthy of note, however, that the common stinging nettle appears to be immune to the toxins contained in the saliva of all three species, and shows no symptoms of attack.

L. pratensis is very common and widely distributed; the same methods of control apply to this species as were recommended against *C. norvegicus*.

NATURAL ENEMIES. Little seems to be known of the parasites of *L. pratensis* in the British Isles; Crosby and Leonard[3] record a Mymarid (*Anagrus ovijentatus*) from the egg of this Capsid.

Calocoris fulvomaculatus De Geer. Needle-nosed Hop Bug, Shy Bug

DESCRIPTION. *Adult.* Male dark olive brown, female paler; sparingly clothed with fine shining golden pubescence; apex of second joint of antennae, centre of the head and two spots on the pronotum black, in the male these spots are hardly perceptible; basal joint of antennae swollen, collar of pronotum slightly raised, sides sinuate; elytra parallel-sided in male, widened behind middle in female; femora mottled with darker brown, posterior pair swollen; tibiae ochreous with short black spines. (Saunders[10].) Length 7 mm. The young stages are greenish to reddish brown with a yellowish green band on the abdomen; the nymphs have dark wing buds.

Egg. Rather shorter and more slender than the egg of *C. norvegicus*, whitish to pale yellow, curved and compressed; cap deep, much compressed laterally with fine longitudinal striations and rounded margin and, as with *norvegicus*, rising slightly higher at the convex than at the concave margin. The neck of the cap is considerably constricted laterally and the egg is gradually contracted towards it. (Steer[13].) Length 1·4 mm.

LIFE HISTORY. The eggs are laid in August and September in hop poles, the larvae hatching in the following spring in May and June and are mature in June, July and August. It is possible that some eggs may hatch in the same year as laid and the young stages pass the winter. There is probably only one generation a year. Much damage is done to the hops by the Capsid puncturing the bine. A dense growth of lateral shoots is formed which renders the plant commercially useless. Here again the damage appears to be mainly due to the toxic action of the salivary secretions. As a control Theobald[14] recommends the use of the ordinary hop wash to which has been added nicotine at the rate of 1 oz. of 98 per cent. nicotine to 10 gallons of wash; this is effective against the young stages but is doubtfully so against the adults. Attacks are worst in hop gardens where poles are used.

HOST PLANTS. Mainly hops, also black currants, various Umbelliferae, sometimes hazel, ash, wild rose, apples and pears.

NATURAL ENEMIES. Butler[2] records an undetermined Proctotrypid, parasitising the nymphs.

REFERENCES

HEMIPTERA—HETEROPTERA

(1) AUSTIN, M. D. (1929). Observations on the eggs of the apple Capsid (*Plesiocoris rugicollis* Fall.) and the common green Capsid (*Lygus pabulinus* Linn.). *Journ. S.E. Agric. Coll.* Wye, Kent. No. 26.

(2) BUTLER, E. A. (1923). *A biology of the British Hemiptera-Heteroptera.* London.

(3) CROSBY, C. R. and LEONARD, M. D. (1914). The tarnished plant bug. *Agric. Exp. Station, Cornell Univ. Bull.* No. 346.

(4) FOX WILSON, G. (1925). The egg of *Lygus pratensis*. *Entom. Mon. Mag.* LXI.

(5) KNIGHT, H. H. (1917). A revision of the genus *Lygus* as it occurs in America, north of Mexico, with biological data on the species from New York. *Agric. Exp. Station, Cornell Univ. Bull.* No. 391.

(6) MOLTZ, E. (1918). Die Wiesenwanze *Lygus pratensis* L. ein Gefähtlicher Kartoffelschädling. *Zeitschr. f. Pflanzenkrank.* XXVII, 7–8.

(7) PETHERBRIDGE, F. R. and THORPE, W. H. (1928). The common green Capsid bug. *Ann. App. Biol.* xv, No. 3.

(8) PETHYBRIDGE, G. H., LAFFERTY, H. A. and RHYNEHART, J. G. (1922). Investigations on flax diseases. *Journ. Dept. Agric. and Tech. Instr. Ireland,* xxii, No. 2.

(9) ROSTRUP, S. and THOMSEN, M. (1923). Bekaempelse af Taeger paa Aebletraeer samt bidrag til Disse Taegers Biologi. (With a summary in English.) *Tidsskrift for Planteavl,* xxix.

(10) SAUNDERS, E. (1892). *The Hemiptera-Heteroptera of the British Islands.* London.

(11) SMITH, KENNETH M. (1920). Investigation of the nature and cause of the damage to plant tissue resulting from the feeding of Capsid bugs. *Ann. App. Biol.* vii, No. 1.

(12) SMITH, KENNETH M. (1925). A note on the egg-laying of *Calocoris bipunctatus* Fab. *Entom. Mon. Mag.* lxi.

(13) STEER W. (1929). The eggs of some Hemiptera-Heteroptera. *Entom. Mon. Mag.* lxv.

(14) THEOBALD, F. V. (1925). *Cultivation, diseases and insect pests of the hop crop.* Ministry of Agric. and Fisheries, Miscell. Publ. No. 42.

CHAPTER VI

HEMIPTERA—HOMOPTERA

Sub-order 2. *Homoptera*

Two pairs of wings usually sloping roof-like over the back when at rest; wingless forms also present. Rostrum arises from the hinder part of the lower side of the head; metamorphosis incomplete, but slightly more developed than in the *Heteroptera*.

Family *Jassidae*. Leaf Hoppers

One of the most abundant forms of the order. They are very active, small and slender insects, usually tapering towards the hinder end. They may be distinguished by the shape of the tibiae which are angular, the hind pair possessing a row of spines on each margin.

Jassids are seldom serious pests of farm crops, though they are always present in greater or lesser numbers upon potatoes. Two species are briefly dealt with, less because of the damage they cause which is slight, than because they may later assume importance as potential vectors of virus diseases. (See Chapter XIV.) Potato leaf hoppers, the only ones considered here, are mostly green or yellow, they are winged in the adult stage. The young forms are wingless and pale in colour and are always found on the under sides of the leaves.

Chlorita viridula Fall.

DESCRIPTION. *Adult.* Green with pale or white markings on the head, pronotum and scutellum; a spot in the apex of the sub-costal and supra-brachial areas, a large sub-triangular spot in the apex of the brachial area, the membrane hyaline, the latter tinged with fuscous; white silky hairs on the apex of the male genital plates with some erect bristles. (Edwards (9).) Length 3 mm.

Eupteryx auratus Liv.

DESCRIPTION. *Adult.* Fore parts yellow or greenish yellow, sides and apex of the frons sometimes black, on the crown two large black spots, which occasionally run together behind; on the pronotum a large black spot on each side which is sometimes kidney-shaped, there is often a pair of black points near the front margin; scutellum with two large black spots at the base; elytra yellow or greenish yellow, with a wide irregular fuscous stripe down the middle, on the costa near the base an oblique black line, and just beyond the middle a large black spot, on the middle of the inner margin a round black spot; abdomen black, hind margins of the segments narrowly yellow, genital plates yellow; legs entirely yellow. (Edwards (9).) Length 4 mm.

LIFE HISTORY. Little is known of the details of the life history of these two leaf hoppers. The writer has bred them easily in captivity on potatoes and there are probably two or three broods in the year. The whole life cycle of *E. auratus* under glasshouse conditions occupies about six weeks. Hibernation is probably in the egg stage, though the adult also may be capable of overwintering, as the writer has found numbers of a closely allied form at the roots of grass tufts in midwinter. In America a somewhat similar species attacking the potato is known to hibernate in the adult stage.

INJURY TO HOST PLANT AND SYMPTOMS OF ATTACK. As with the Capsid bugs the saliva of these insects is toxic to the plant, each puncture being marked by a white spot where the chlorophyll and cell contents have been destroyed. (Smith, *Ann. App. Biol.* XIII, 1926.) Jassid attack can always be diagnosed from these characteristic white spots which coalesce in a bad attack and form a bleached area.

HOST PLANTS. These hoppers attack a very large number of plants, but the potatoes are the only agricultural crop likely to suffer. The nettle is the favourite wild host of *E. auratus*.

DISTRIBUTION. Common all over Great Britain, especially in the eastern counties. The common potato leaf hopper in Ireland is probably a different species.

CONTROL. A soap and nicotine sulphate spray does not appear to control leaf hoppers, probably on account of their great activity, and the consequent difficulty of hitting the insects with the spray. Four years' experience of spraying for potato leaf hoppers in America has shown this spray to be unsatisfactory as it only kills the young forms. Bordeaux mixture (4 : 4 : 50) appears to be a good remedy, probably acting as a deterrent. The ordinary routine spraying of potatoes with Bordeaux mixture will probably suffice to control these insects, which rarely demand special measures in the British Isles.

Family *Aleyrodidae* (*Aleurodidae*). White Flies; Snow Flies

Very small, moth-like insects, wings whitish, opaque; body and wings covered with a white meal; antennae 6-segmented, second segment long; there is a quiescent pupal stage.

This is a small group of insects of little agricultural importance. There is one species however, *Aleurodes brassicae*, which is a pest of cabbages and other Brassicas and is dealt with shortly.

Aleurodes brassicae Walk. The Cabbage White Fly

DESCRIPTION. *Adult.* In the female the head, eyes and thorax are black, with some yellow on the latter; abdomen yellow, somewhat cylindrical at base, the whole covered with a waxy coating; antennae 6-jointed, the first joint swollen. There is a blackish mark across the centre of the fore wings, which appears as a black bar across the insect when the wings are closed; apices of wings rounded. Length 2 mm.

Egg. Long-oval in shape, standing upright, firmly fixed to the epidermis of the leaf; shining, pale, translucent, later becoming dark in colour; covered with meal.

Larva. Motile on hatching from the egg, soon becomes fixed and scale-like, white with two yellow spots.

Pupa. Immobile, pale with red eyes.

LIFE HISTORY. The eggs are laid on the leaves of Brassicas, they are deposited on the under sides in patches of white waxy material in clusters of three or more embedded or cemented firmly in the tissue. They hatch in about 12 days and after 10 days the larvae enter the quiescent pupal stage, from which the adult emerges in about 4 days. In the warmer parts of England *A. brassicae* seems capable of wintering out of doors, continuing to breed all the year.

HOST PLANTS. Mostly Brassicas, both wild and cultivated. The writer has found it on several occasions breeding on the dandelion (*Taraxacum*), but these have usually been in the vicinity of cabbage fields.

INJURY TO HOST PLANT AND SYMPTOMS OF ATTACK. Loss of vitality is caused by the draining of the sap due to the sucking of the white flies. Infested cabbages are sometimes withered but are more often marked with whitish or yellowish patches. The insects feed on the under surface of the leaf from which they arise in a white cloud when disturbed, returning immediately afterwards to the plant.

DISTRIBUTION. Rare in the north and midlands, but more common in the south and south-west, particularly Devonshire, where it was a serious pest in 1925–6 and in Cambridgeshire in 1929.

CONTROL. No good method of control has been devised for *A. brassicae*. The only recommendations seem to be the destruction by burning of infested leaves or spraying several times with a solution of soap and water.

NATURAL ENEMIES. There appears to be no record of the natural enemies of *A. brassicae* in Great Britain, but it is possible that two important Chalcid parasites of the closely allied greenhouse white-fly (*Trialeurodes vaporariorum*), i.e. *Encarsia partenopea* Masi. and *E. formosa* Gahan, may also attack the former species. Mercet(15), in Spain, records the Hymenopterous parasite *Prospaltella conjugata* Masi. from *A. brassicae*.

Family *Aphididae*. Green Flies; Plant Lice

Body somewhat pear-shaped, legs long and slender, not adapted for leaping; antennae 3–6-jointed; two pairs of transparent wings which are often absent; a pair of tubes or cornicles usually present on the dorsal surface of the fifth abdominal segment. Metamorphosis incomplete.

Aphides, as a rule, have a somewhat complicated life history and show considerable dimorphism in form and colour between individual insects of the same species. *Parthenogenesis* is a characteristic of this group of insects and the females may be either viviparous or oviparous. Many aphides have a definite cycle of movements from host to host during their life history, these hosts often belonging to widely differing orders of plants. As a group, aphides are small insects, mostly green or brown in colour, usually from 2–3 mm. long and rarely exceeding 5 mm. in length. They are sluggish, slow-moving creatures, but are exceedingly prolific and constitute some of the worst insect pests of plants.

The classification and nomenclature of this immense group is a matter of great complexity and the writer has only attempted to give complete accounts of a few of the more important aphides; for more information on the other species which are only shortly dealt with here, the reader is referred to the comprehensive work of F. V. Theobald, *The Aphididae of Great Britain*, published in three volumes, and to Buckton's *Monograph of British Aphides* (Ray Society).

Genus *Macrosiphum* Passerini

Characters of genus after Theobald: frontal tubercles very prominent and markedly divergent; antennae long, as long as or longer than the body of six segments; flagellum long, third segment longer than the fourth with circular sensoria in alatae and sometimes in the apterae, and in the alatae with sensoria now and then present on the fourth segment; cornicles long, cylindrical, rather thin and often tapering towards the apex; cauda long, lanceolate or sword-shaped, often constricted near the middle; wings normal; legs usually long and slender. Males alate; oviparous females apterous.

Macrosiphum avenae F. (*granarium* Kirby.) The Grain Aphis
SYNONYMS. *Macrosiphum cerealis* Kalt.; *M. avenivorum* Kirk; *Aphis hordei* Kyber; *A. avenae* Walk.

DESCRIPTION. A green species, cornicles black, cauda pale green in apterous forms, yellow in alate forms.
Apterous female. Colour yellowish green sometimes darker towards hinder end of body; head yellow to brownish yellow; eyes red; antennae usually black or dark-coloured except at the base; terminal half of the femora, apex of the tibiae and tarsi dark, rest of legs yellow. The body is frequently marked with a brownish spot on each side of the prothorax with occasionally a row of dark spots on each side of the abdomen and between the cornicles; cauda long, stout and curved upwards, about two thirds the length of the cornicles; cornicles black and rather long with the tips reticulate. Length 2·4–2·8 mm.
Winged viviparous female. Colour yellowish green to green; head brown; eyes red or brown; antennae black, the first joint sometimes lighter; cornicles black, tips reticulate; cauda yellowish; abdomen marked with four or five small transverse blackish spots and four black lateral spots in front of the cornicles; legs similar to those of the apterous female; antennae longer than body; wings about twice the length of body. Length 1·4–2·6 mm. (Pergande (18).)

LIFE HISTORY. The egg is very small, elliptical and pale in colour when first deposited, later turning black. The eggs are laid in the autumn on grasses and self-sown corn, the agamic females, or stem mothers, hatch from these overwintering eggs in March and reproduce asexually. About June the winged females are produced which leave the grasses and fly to cereal crops. Here they remain throughout the summer, producing asexual generations, feeding on the ear and blades of the corn. About harvest time, when the plant tissues and ears of corn become tough, winged females leave the corn and return to wild grasses where the oviparous females are produced. Later, males appear on the corn and migrate to the wild grasses where they fertilise the oviparous females. The winter eggs are then laid and the life cycle completed.

HOST PLANTS. Wheat, barley and oats and a number of wild grasses, among which the following may be mentioned: meadow foxtail (*Alopecurus pratensis*), wild oat (*Avena fatua*), brome (*Bromus* spp.), cock's foot (*Dactylis glomerata*), sheep's fescue (*Festuca ovina duriuscula*), soft holcus (*Holcus mollis*), meadow poa (*Poa pratensis*), common persicaria (*Polygonum persicaria*).

INJURY TO HOST PLANT. The plants lose vitality and become weakly owing to the draining of the sap. The aphides do not usually become noticeable until the corn is well advanced; a bad attack of this and allied species will very materially reduce the yield in both quantity and quality.

DISTRIBUTION. Common all over Great Britain and Europe, it occurs also in America and Japan. An attack on oats was recorded from Yorkshire in 1929.

CONTROL. Little can be done in the way of control for this aphis; the destruction of the winter hosts in the shape of wild grasses and self-sown corn, especially on headlands and along hedgerows surrounding the corn fields, is a useful measure.

NATURAL ENEMIES. Theobald records the following natural enemies in Great Britain: the Braconid *Ephedrus* sp.; Coleoptera, *Adalia* (*Coccinella*) *bipunctata*; and among Diptera various species of Syrphids. Attempts have been made to control this aphis by artificially introduced infestations of *Aphidius granarius* Marsh, under field conditions (Arthur (1945), *Bull. Entom. Res.* XXXVI, 291–5). In other countries the following are recorded: in Russia, the Braconid *Aphidius* sp. and in the United States of America the Chalcid *Aphelinus semiflavus* How., and among Diptera the predacious larvae of the Cecidomyid *Aphidoletes meridionalis* Felt.

Macrosiphum gei Koch. (*solanifolii* Ashm.). The Pink and Green Potato and Rose Aphis

DESCRIPTION. General colour green, head darker with a narrow dark border along the hinder margin, thorax dark green to brownish, abdomen yellowish green, ventral surface light green, cornicles long, green and slightly tapering. The following descriptions of the winged viviparous female and the apterous viviparous female are quoted from the work of Davidson (3).

Winged viviparous female. Head somewhat narrow with many finely capitate hairs; eyes black; ocelli well developed; antennae with the two basal segments pale green, remainder dark or fuscous, as long as or longer than the body; segment 3 with about 14 sensoria in a row along its internal face, extending for about three fourths the length of the segment; several small, faintly capitate hairs on all segments. Thorax; wings well-defined, dark veins and greenish brown stigma. Legs; coxae and trochanters green, femora and tibiae somewhat darker green and darker at distal ends, tarsi dark. Abdomen elongate with rows of short hairs on all segments; cauda not quite half the length of cornicles, pale green, ensiform, finely spinose and bearing several lateral hairs, and one or two subapical median hairs; anal plate rounded, spinose and bearing a few long hairs; cornicles yellowish green basally, darker on distal portion, slender, tapering distally, reticulate at distal end for about one fifth of its length and the surface sculptured over remaining portion.

Apterous viviparous female. Body green to yellowish green, more or less uniformly yellowish green on the venter. Head green, bearing a few short hairs; eyes black; rostrum yellowish green, darker at distal end, reaching to coxae 2; antennae as long as or longer than the body, fuscous, apex of segments darker and also the whole of segment 6; segment 3 with a few (about three) round sensoria on inner face near its base, all segments with a few faintly capitate hairs. Thorax; legs yellowish green with dark tarsi and distal end of tibiae and femora dark. Abdomen elongate with rows of short hairs on all segments; cauda slightly less than half the length of cornicles, pale green, ensiform, finely spinose, bearing several lateral hairs and one or two sub-apical median hairs; anal plate as in the winged viviparous female; cornicles pale green basally, darker on distal portion, about as long as antennae segment 3, reticulate at distal end for about one fifth of the length. (Fig. 6.)

LIFE HISTORY. In America, where this aphis is a more serious pest than in Great Britain, the winter is passed in the egg stage upon rose trees. These hatch in the spring and after the production of some generations on the rose, a migration takes place of both winged and wingless forms to the potato where it multiplies exceedingly. In this country there is probably a similar sequence of events, but in addition, there is some evidence that the aphis may winter in the egg stage upon sprouting 'seed' potatoes, and it also breeds asexually all the year round in glasshouses. It has a wide range of host plants and in addition to attacking the potato which gives this aphis its agricultural significance it may be found during

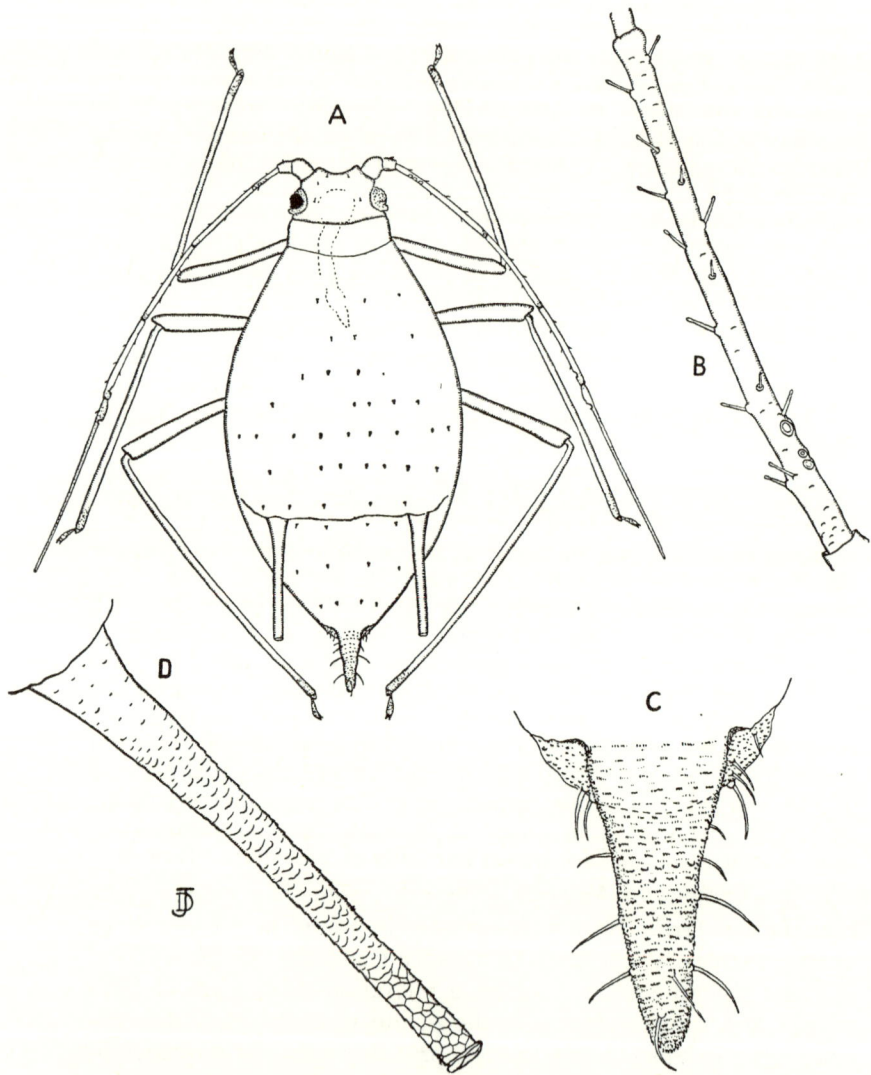

Fig. 6. *Macrosiphum gei* Koch., from tulips: *A*, apterous viviparous female; *B*, segment 3 of antenna; *C*, cauda and anal plate (dorsal view); *D*, cornicle, greatly enlarged. (After Davidson.)

the summer upon the iris, tulip and nettle. Although not as a rule
a severe pest of potatoes, it does occur upon them occasionally in
sufficient numbers to cause material damage. Its chief significance,
however, lies in the fact that it is suspected of being a potential
vector of the virus diseases of the potato. (Chapter xiv.) Until
further research has proved the contrary, some suspicion rests on
all the aphides attacking the potato.

HOST PLANTS. Among cultivated plants the most important are
the potato, iris, tulip and rose. Theobald has recorded it upon hops.
Among weeds, it often occurs, like most of the potato insect fauna,
upon nettles. Davidson [7] gives the following list of host plants
for this species: *Rosa* spp.; *Solanum*; *Aster*; Brassicae; *Chenopodium* sp.; willow herb (*Epilobium* spp.); lettuce (*Lactuca* sp.);
Phaseolus; sow thistle (*Sonchus* spp.). Patch [16], in America, gives
a list of 84 food plants belonging to 32 different orders.

INJURY TO HOST PLANTS AND SYMPTOMS OF ATTACK. The aphis
crowds upon the tender growing points and terminal leaves of the
host causing them severe injury. The leaves become deformed and
dry, the edges turn downwards and are frequently yellow in colour.
Colonies of the aphides also establish themselves upon the under
sides of the lower leaves and upon the blossom.

DISTRIBUTION. Common generally throughout the British Isles;
Davidson [7] gives the following localities: Devon, Essex, Hertfordshire, Isle of Wight, Kent, Lincolnshire, North Wales, Sussex,
Yorkshire, Westmorland, Scotland, Ireland. The writer has found
it in Lancashire, Cheshire, Derbyshire and Cambridgeshire.

CONTROL. A good combination spray for potatoes consists of
8–10 oz. of nicotine sulphate added to 50 gallons of ordinary
Bordeaux mixture. Two or three sprayings should be given at
intervals of about 10–14 days, the first when the plants are 5–6
weeks old. This is a useful spray which is effective against most
of the pests of the potato. Some caution is needed, however, in
the use of Bordeaux mixture in the presence of very heavy aphis
infestation, as under these circumstances it has been known to
cause scorching of the foliage.

NATURAL ENEMIES. There are few records of the parasites of
M. gei in Great Britain; the writer has some evidence that the
Braconid *Aphidius matricariae* Hal. attacks this aphis. The follow-
ing are recorded from *M. gei* in America: the Braconids *Aphidius*

polygonaphis Fitch. and *A. rapae* Curt., the Chalcids *Aphelinus jucundus* Gahan, *Pachyneuron aphidivorum* Ashm. and the Proctotrypid *Lygocerus* sp.

Macrosiphum pisi Kalt. Pea Aphis; Green Pea Louse

DESCRIPTION. *Apterous viviparous female.* Body green to pale green, shining, usually darkening towards posterior; eyes red or reddish brown; antennae long arising from prominent tubercles; apices of segments dark, segment 6 dark brown and the longest, segment 3 with 1–3 circular sensoria near the base; rostrum reaches as far as coxae of second pair of legs; legs long and slender, green, apices of femora and tibiae dusky, tarsi black; cornicles long and slender, broadest at base, non-reticulate; cauda green, long and ensiform with a few long hairs. Length 2·5–3 mm.

Winged viviparous female. Green; head sometimes yellowish; eyes red; antennae very long reaching beyond cauda, pale, segments 1 and 2 darker green, segment 6 black, segment 3 with a row of 10–20 sensoria; rostrum shorter than in the apterous viviparous female, wings clear, veins brownish, stigma yellowish; legs green, darkened as in the apterous viviparous female, cornicles long and thin, reaching beyond cauda, non-reticulate, apex darker, imbricated, cauda long and green. Length 2·79 mm. (Fig. 7.)

LIFE HISTORY. Unlike most aphides the life of this species is spent mostly upon plants of the same natural order (Leguminosae), peas, clover, vetch, lucerne, etc. A certain degree of migratory movement takes place among these various plants. The winter is passed in the egg stage, and occasionally in mild winters as apterous females upon perennial Leguminosae such as clover and everlasting pea. The aphis also breeds throughout the summer on these hosts. In June and July, numbers appear on field and garden peas and beans where they breed until August and September, when winged females are produced which migrate to the clover. Interchange of the colonies upon clover and peas may also occur during the summer. This aphis is exceedingly prolific, as many as 4–13 young being born daily on blossoming peas and 3–10 on clover. The sexuales do not occur commonly in this country, and are only produced in autumn and early winter. Theobald describes the eggs as usually deposited low down on the haulm of everlasting pea, but they are occasionally found on other parts of the plant.

HOST PLANTS. Mostly leguminous plants, cultivated peas both culinary and ornamental; wild everlasting pea (*Lathyrus sylvestris*), red clover (*Trifolium pratense*); wild white clover (*T. repens*); alsike (*T. hybridum*); broad beans (*Vicia faba*); lucerne, vetches, wild and cultivated broom (*Cytisus*). Theobald also records it from green-weed (*Genista tinctoria*) and shepherd's purse (*Capsella bursa-pastoris*).

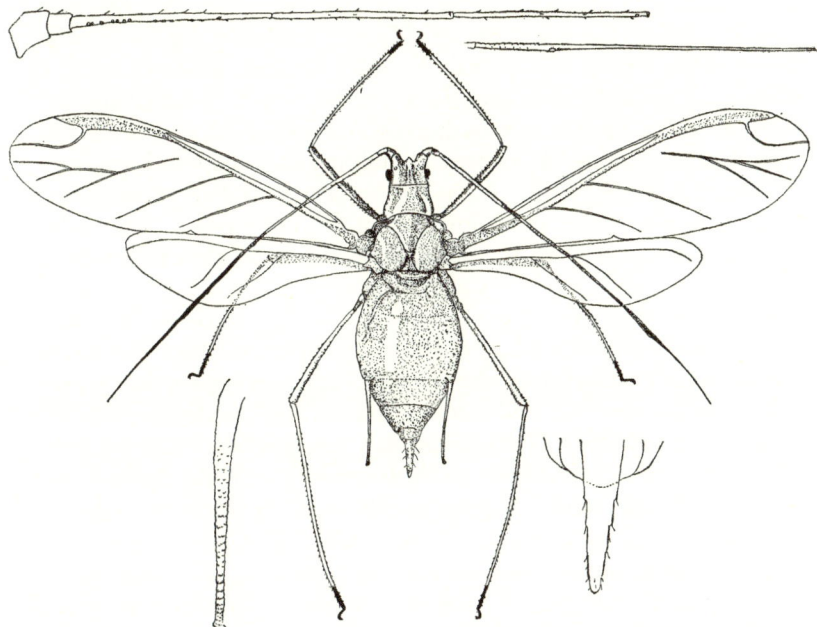

Fig. 7. *Macrosiphum pisi* Kalt. The pea aphis. Winged viviparous female. *Above*, antennae; *below, right*, cauda; *left*, cornicle. (After Davis.) Much enlarged.

INJURY TO HOST PLANT AND SYMPTOMS OF ATTACK. An attack by *M. pisi* upon peas is followed by stunting of the plant, yellowing and malformation of the leaves and pods and even by the death of the plant in severe cases. Serious damage can be caused to peas by this aphis, especially to late-sown crops. In America it is suspected of being the vector of a virus disease between clover (*Trifolium*) and lucerne (*Medicago*), it also transmits pea mosaic.

DISTRIBUTION. Common in many parts of the British Isles, especially in the pea-growing centres of Lincolnshire and Cambridgeshire; it also occurs in Europe, America where it is a serious pest, Africa, India and Japan. In this country recent severe attacks have been reported from south Lincolnshire, north Bedfordshire, Huntingdonshire, Cambridgeshire, Norfolk and Kent.

CONTROL. The attacks of this aphis are largely influenced by climatic conditions. Warm wet weather is very favourable to the development of the aphides on peas, while cold weather, cold rains, dry winds and drought delay their development. Heavy rains greatly reduce the numbers of *M. pisi*, probably by mechanical dislodgement from the plant. Early-sown peas are less liable to damage. This aphis has been found to be highly resistant to nicotine and, if it is used, a dust with a high nicotine content is advisable. Theobald recommends a spray containing soft soap and quassia.

Glasgow (1946, *Journ. Econ. Entom.* XXXIX, 195–9) recommends D.D.T. applied as an emulsion against the pea aphis, a 50 per cent. benzol solution of D.D.T. in a weak bordeaux or some other cheap emulsifying agent.

NATURAL ENEMIES. The larvae of the Syrphid *Lasiophthicus* (*Catabomba*) *pyrastri* L. and the Coccinellid beetles *Adalia bipunctata* L., *Coccinella septempunctata* L., *C. decempunctata* L. var. *variabilis* Ill. and *C. undecimpunctata* L. are predacious on this aphis. The following Hymenopterous parasites are recorded from *M. pisi* in America: the Braconids *Aphidius rosae* Hal., *Aphidius* sp., *Praon alaskensis* Ashm., *P. simulans* Prov., *Lysiphlebus* sp., *Megorismus fletcheri* Crawf. and the Proctotrypid *Lygocerus niger* How. Fluke(11) records 76 parasites and predators of *M. pisi* in North America, of which the majority are Syrphids. In Slovakia an outbreak of the pea aphis on lucerne was largely controlled by the entomogenous fungus *Entomophthora aphidis* Hoffm.

Genus *Myzus* Passerini

Theobald (27) gives the characters of this genus as follows: Frontal tubercles distinct especially in apterae, they project inwards and are strongly gibbous, notably in the males; antennae of six segments, the first gibbous, like the frontal processes, about equal in length to body; cornicles rather long, cylindrical, in some slightly swollen in places; cauda prominent, somewhat conical, constricted very slightly or not at all, from half to less the length of the cornicles; legs moderately long; wing venation normal; oviparous females wingless; males winged.

Myzus persicae Sulz. Potato and Peach Aphis;
Spinach Aphis of America

DESCRIPTION. *Apterous viviparous female.* Rather small, ovate, shiny, of various shades of green, yellow or rose; head broad, frontal tubercles prominent, antennae greenish and slightly shorter than the body; legs greenish, apices of tibiae and tarsi dark; cornicles slightly swollen, tips dusky. This species varies greatly in colour and differs somewhat in appearance according to its host plant, specimens from spinach being paler and more shiny than those from cabbage or potato. Length 2–2·5 mm. (Fig. 8.)

Alate viviparous female. Head and thorax black, abdomen greenish with one or two transverse dark bands, a large dark patch between the cornicles and four lateral dark spots; cornicles dark and slightly swollen towards the middle, cauda dark and pointed; legs pale yellow, greenish or reddish, femora and apices of tibiae dark, tarsi black; third joint of antennae tuberculate with a number of sensoria; wings rather large. Length 2–2·5 mm.

For a description of the oviparous female and the male, see Theobald, *Aphididae of Great Britain*, I, 321–2.

LIFE HISTORY. *M. persicae* is a cosmopolitan and practically omnivorous insect, feeding and breeding upon an enormous variety of plants. It can be found at any time of the year as it lives all through the winter under glass, breeding asexually; it thrives best between the temperatures of 60° and 80° F. Normally it lays its eggs in autumn on peach, nectarine, and *Daphne*; and in America upon plums, cherry and peach. The eggs may hatch as early as January and occasionally the adults are to be found during the winter in protected places upon the stems of the summer food plant; the writer has known it to pass a mild winter in an unheated glasshouse upon various Brassicae. In the south and south-west of the United States of America and in South Africa *M. persicae* partly loses its migratory habit and feeds all the year round upon low-growing vegetables and weeds. In the north of the United States and in this country it normally hibernates in the egg stage upon the stone fruits above mentioned. The life of an individual female of *M. persicae* averages about 36 days, the approximate number of young produced being about 40.

HOST PLANTS. Common upon potatoes, of which plant it is often a severe pest. This insect with *Macrosiphum gei* and *Myzus pseudosolani* are the commonest potato aphides, and all three are found infesting sprouting potato tubers in store; *M. persicae* often attacks cabbages and other Brassicas. Among the many additional host plants may be mentioned garden plants of all kinds, sugar-beet, woody shrubs such as *Daphne*, peach and nectarine, the common stinging nettle (*Urtica dioica*) and the Solanaceous weeds such

52

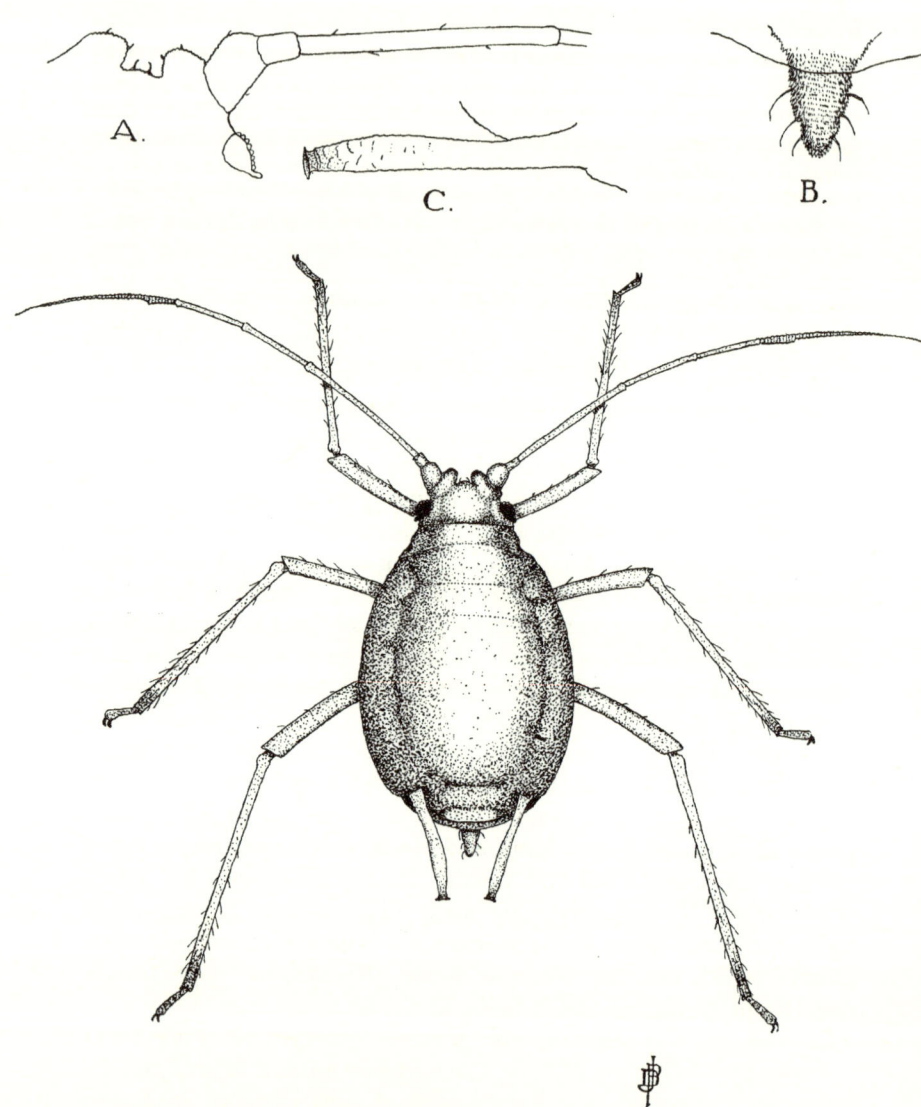

A.

C.

B.

Fig. 8. *Myzus persicae* Sulz. Apterous viviparous female. × 27. A = head and
antenna; B = cauda; C = cornicle. × 90.

as bittersweet (*Solanum dulcamara*) and the black nightshade (*Solanum nigrum*). The particular significance attaching to this last plant and *M. persicae* is explained in Chapter XIV on virus diseases. Davidson (*List of British Aphides*, p. 60) gives a list of 52 plant hosts for this aphis.

INJURY TO HOST PLANT. Much damage is occasionally caused to potatoes by heavy infestations of *M. persicae* on the haulm; a more insidious mode of attack, however, is that on the sprouting tuber in store. As *M. persicae* has been proved to be a very efficient vector of the serious potato virus disease leaf roll, as well as other potato viruses (Smith(22)), this attack on the sprouts is an important matter. (See Chapter XIV.) The affinity which *M. persicae* appears to possess for plant virus diseases of many kinds, reinforced by its worldwide distribution and omnivorous habits, renders it an insect of the gravest importance.

DISTRIBUTION. All over Europe, North and South America; North and South Africa; Australia; New Zealand; Japan; India; Iraq; Bermuda.

CONTROL. The same spray as that recommended against other Hemipterous insects on potato may be used against this aphis. Potatoes in store should be regularly fumigated with a nicotine fumigant. Dusting the sprouting tubers with a 5 per cent. nicotine sulphate dust is also recommended.

NATURAL ENEMIES. Like most aphides this species is often heavily parasitised by Braconids. Under glass the writer has found the commonest parasite to be a species of *Aphidius*, probably *A. matricariae* Hal. which may exert a high degree of control under glasshouse conditions. Sometimes accompanying *Aphidius* is the wingless Cynipid *Nephycta pedestris* Curtis (Waterston det.). The following Hymenoptera are recorded from *M. persicae* in America: the Braconids *Aphidius phorodontis* Ashm., *Diacretus rapae* Cust., *Ephedrus incompletus* Prov., *Praon simulans* Prov. and the Chalcid *Aphelinus semiflavus* How. In Hawaii the Braconid *Diacretus chenopodiaphidis* Ashm. is recorded. According to Nolla (*Phytopathology*, XIX, 1, 1929), some measure of control may be effected by the artificial dissemination of the entomogenous fungus *Acrostalagmus aphidium*. A spore suspension was sprayed on a field of egg-plants (*Solanum melongena*) heavily infested with *M. persicae*; under favourable conditions of humidity this appeared to give good control.

Myzus pseudosolani Theob.

DESCRIPTION. Very similar to *M. persicae*, but it can be differentiated by its pale cylindrical cornicles, and in the apterous female by the presence of one or two sensoria at the base of the third antennal segment. The frontal tubercles are less pronounced than in *M. persicae* and the apices of the cornicles more widely flared and darker at the tips. It differs from *Macrosiphum gei* in the non-reticulate cornicles. In the winged female, also, the banding of the abdomen is sometimes very marked. Theobald gives a detailed description of this species in *Aphididae of Great Britain*, I, 313.

LIFE HISTORY. The life history of this aphis has been studied by Patch (17). The overwintering eggs are deposited on the garden foxglove (*Digitalis purpurea*). These hatch in spring and the resulting stem mothers shelter in the folded leaves; later, winged females are produced and these migrate to potato and other plants from June onwards, returning to the foxglove in September and October. *M. pseudosolani* is especially abundant upon sprouting potato tubers in company with *M. persicae*. It occurs from January to April upon the 'seed' potatoes and on the haulm from July to October. By its feeding this aphis causes much curling of the leaves and twisting of the petioles. It is also an occasional vector of potato leaf roll.

Myzus festucae Theob.

DESCRIPTION. In the *alate viviparous female* the head and thorax are brown, abdomen green with two rows of dark spots; cornicles green, cylindrical; cauda green, much thicker than cornicles; legs green. Length 1·8–2 mm.

The *apterous viviparous female* is pale green or greenish pink; antennae longer than the body, green; cornicles green, cylindrical, rather thin; cauda green, rather thick and bluntly pointed; legs green, as body. Length 1·5–2 mm. (See Theobald, *Aphididae of Great Britain*, I, 335.)

This insect is sometimes a pest of wheat, barley and oats. The writer has seen it occasionally swarming in the pasture fields in Cheshire where it was killing the grass and affecting the health of the stock which swallowed quantities of the aphis. Later in the season, large numbers of parasites appeared which cleared off the attack. *M. festucae* occurs on the following grasses in addition to others: fescue grass (*Festuca ovina*), oat grass (*Avena* sp.), and brome (*Bromus* spp.).

Myzus lactucae Schrank

This aphis migrates in May and June from currants and gooseberries which are the winter hosts, to lettuces where it feeds upon the heart and may do much damage. On lettuce the apterous female is yellowish to light green, the abdomen is marked with seven pairs of dark lateral spots, cornicles yellowish, dusky at apices, cylindrical; cauda yellow; head and thorax dark. (Theobald, *op. cit.* 1, 302.)

Genus *Aphis* Linn.

Head without any prominent frontal tubercles; antennae of six segments sensoria round; cornicles cylindrical to slightly tapering; cauda usually not so long as cornicles, rather elongate or subconical, in most constricted near the middle; wings usually normal. Males alate; oviparous females apterous. (Theobald.)

Aphis fabae Linn. The Bean Aphis; 'Black Fly';
'Black Dolphin'; 'Blight'.

DESCRIPTION. *Apterous viviparous female.* Body elongate oval, colour variable, black to olive green, often with irregular darker pigmented areas over the abdomen, small hairs scattered over the body. Head, eyes black with prominent accessory eyes; antennae 6-jointed; segment 1, apical portion of segment 5 and proximal portion of segment 6, black, remainder paler, . . . a single sub-apical sensorium on segment 5, a compound sensorium on segment 6. Thorax with a prominent prothoracic tubercle on each side; legs black, tibiae and proximal portion of femora paler; segments bearing stout hairs, especially the tibiae. Abdomen with two lateral tubercles on each side, and sometimes one or two small indefinite tubercles; cornicles black, tubular, imbricated, slightly tapering distally, usually one and one third to one and a half times the length of the cauda; cauda with distal half slightly spoon-shaped, black, clothed with short stout bristles and several long curved hairs. Length 2·5 mm. (Fig. 9.)

Winged viviparous female. Body: head and thorax black to brownish black, abdomen varying from brownish black to olive-green, usually with irregular darker pigmented areas, small hairs scattered over the body. Head black, eyes black with prominent accessory eyes, antennae dirty brown to black, varying in length, about two thirds length of body, rostrum dark toward distal end. Thorax with two prominent lateral tubercles on prothorax; wings normal; legs somewhat longer than in the apterous viviparous female, otherwise similar. Abdomen varying in colour from dark velvet black to olive-green, usually with five irregular pigmented areas along the lateral dorsal area and irregular transverse areas segmentally arranged, lateral tubercles prominent; cornicles black, tubular and slightly tapering; cauda not so large as in the apterous viviparous female. Length 2·4 mm. (Davidson(6).) (Fig. 10.)

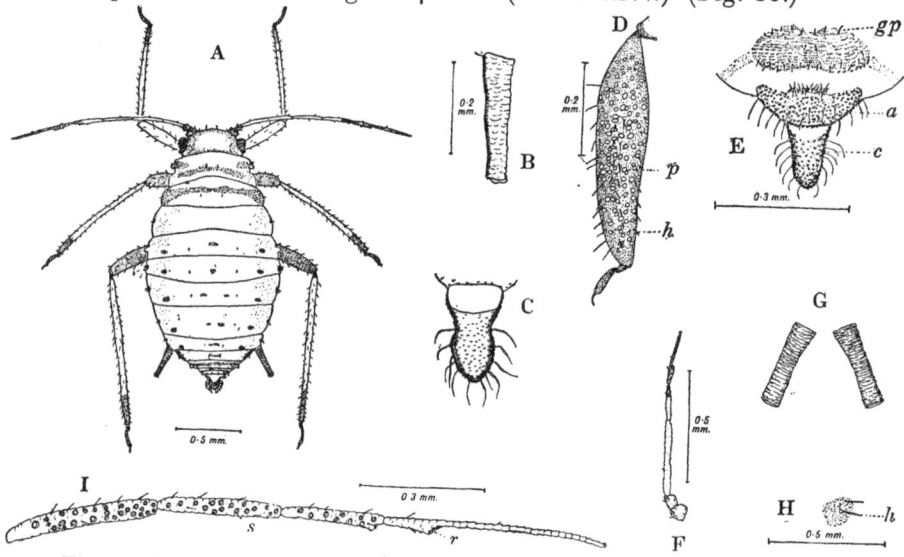

Fig. 9. *Aphis* (*Doralis*) *fabae* Scop. A–C, wingless viviparous female; D, oviparous female; E–H, fundatrix; I, male. A, dorsal view of the wingless viviparous female; B, cornicle; C, cauda; D, flattened hind tibia; E, ventral view of posterior end; F, antenna; G, cornicles; H, marginal area of an abdominal segment; I, antenna. *a*, anal plate; *c*, cauda; *gp*, genital plate; *h* and *h*, hair; *p*, pseudo-sensorium; *s*, secondary sensorium; *r*, primary sensorium. (After M. G. Jones.)

LIFE HISTORY. This very common aphis has a wide range of host plants belonging to many different orders. The normal life cycle is spent partly upon woody and partly upon herbaceous plants, though it appears possible for the insect to live asexually throughout the year upon herbaceous plants alone. The winter eggs are deposited on the spindle tree (*Euonymus*) and the sterile guelder rose (*Viburnum*) during October, near the base of the leaf buds or in crevices of the branches, and hatch out in April about the time of the appearance of the young leaves. The stem mothers which arise from the winter eggs produce a brood of wingless viviparous females which, in their turn, produce a race of winged females and these migrate in May and June from the spindle tree to herbaceous plants such as the broad bean, mangold, etc., and to docks and other weeds. The summer is spent on these herbaceous plants, where many generations of aphides are produced, the host plant being frequently smothered with the insects. During the

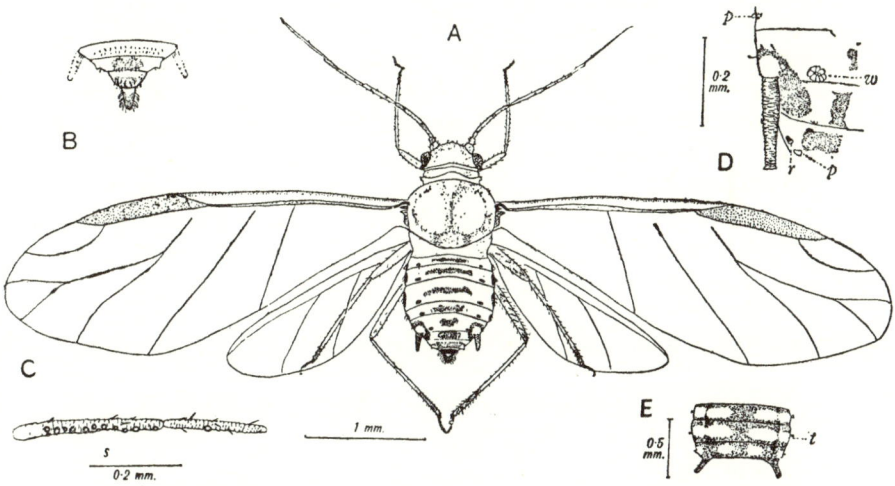

Fig. 10. *Aphis (Doralis) fabae* Scop. A–D, winged viviparous female; E, winged nymph. A, dorsal view of winged viviparous female; B, ventral view of posterior end; C, 3rd and 4th antennal joints; D, cornicle and its surrounding markings; E, dorsal view of 3rd, 4th and 5th abdominal segments to show white wax markings. *p*, papilla; *r*, spiracle; *s*, secondary sensorium; *t*, transverse white band; *w*, wax gland. (After M. G. Jones.)

summer winged forms are produced which migrate to other herbaceous plants, either of the same or different species. In September the sexual individuals are produced on the summer food plants; these are winged, ready for the return migration to the spindle tree where the fertilised winter eggs are deposited. In addition to this life cycle on woody and herbaceous plants, it is possible that the aphis completes its life history on herbaceous plants only, laying its eggs on the dock (*Rumex*), the white goose-foot (*Chenopodium album*), or the broad bean, and migrating from there to other herbaceous plants. As already mentioned, the aphis may in addition continue to breed asexually all the year round in sheltered places on herbaceous plants. Horsfall[13] suggests that the markedly polyphagous habit of *A. fabae* may result in the formation of biological races which in time will exhibit distinctive morphological characters, and which will be confined to certain food plants or sets of food plants. Fig. 11 is a diagram showing the life cycle and different forms of *A. fabae*. For further information concerning this aphis the reader is referred to the work of Davidson, who has made a special study of the insect.

HOST PLANTS. Spindle tree (*Euonymus*), bean (*Vicia faba*), dock (*Rumex*), spinach, parsnip, beet, sugar-beet, mangold, thistle (*Carduus* spp.), shepherd's purse (*Capsella bursa-pastoris*), white goosefoot (*Chenopodium album*), and *Dahlia* sp. For complete list of food plants see Davidson, *List of British Aphides*, p. 76.

INJURY TO HOST PLANTS. Often whole fields of beans are killed by the swarms of this aphis. The flowers of broad beans smothered with the black aphides are a common enough sight in this country. On mangolds and docks some leaf curling is caused. Much secondary injury to the plant arises from the copious excretion of honeydew, which covers the plant with a sticky coating which scorches the leaves and encourages the growth of fungi.

DISTRIBUTION. Common and widely distributed in this country, Europe generally, the United States of America, Canada, Africa, India and Japan.

CONTROL. Temperature influences the development of *A. fabae* to a considerable extent. Cold weather following upon a warm spell in spring often kills large numbers of the newly hatched aphides. Development occurs most rapidly at a mean temperature of about 70° F. Many of them are destroyed by heavy rain. Davidson[4] has shown that certain varieties of field beans are less liable to attack than others. Taking Longpod as representing 100 per cent. of susceptibility 98 per cent. are Bohus Bean (Sweden) and Mazagan; 55 per cent. Heligoland and Granton (Scotch variety), 39 per cent. Spring Tick (Suffolk), winter beans, Small Tick, Carse, Kilbride; while *Vicia narbonensis*, which is thought to be the prototype of the bean, showed only 3 per cent. susceptibility. Early sowing is advisable in order that the plants may be well advanced before the aphides appear, and watch should be kept during May for the early stages of attack, when any heavily infested plants noticed should be removed. It is not advisable to grow sugar beet or mangolds, which are susceptible crops, in the vicinity of beans, which are so often

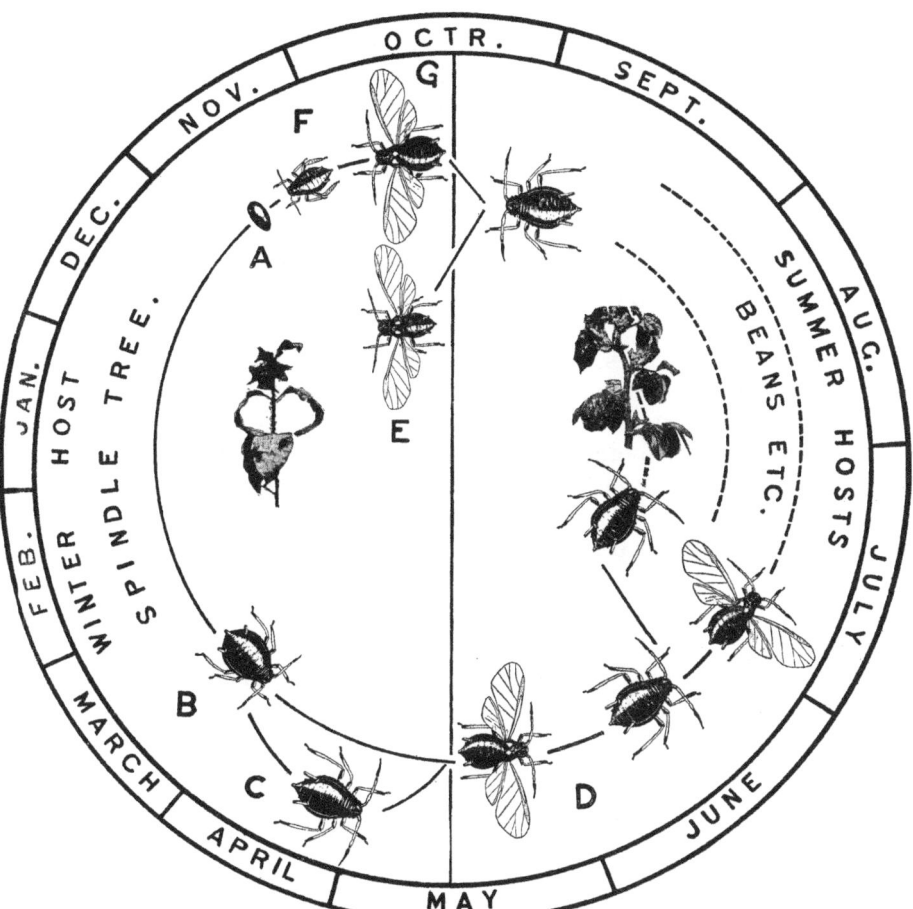

Fig. 11. Diagram illustrating the life cycle of the bean aphis (*Aphis fabae*) and showing the different forms through which the insect passes. The period spent on the winter host is represented on the left half of the circle; that on the summer hosts on the right half. The sectors of the circle indicate the months and the concentric dotted lines a varying number of summer generations. *A*, egg; *B*, fundatrix or stem mother; *C*, wingless viviparous female; *D*, winged viviparous female (migrant); *E*, male; *F*, sexual egg-laying female; *G*, autumn remigrants which produce the sexual females. (After Davidson.)

affected. The removal of the spindle tree which is the winter host of *A. fabae* is unlikely to have a very beneficial effect as the aphis is capable of wintering on other plants. Such weeds as shepherd's purse, goosefoot, docks and poppies should be eradicated wherever possible in the vicinity of susceptible crops. In gardens and allotments, the young growing tips of the plants should be pinched out to prevent the downward spread of the infestation. If a spray is used, one containing nicotine and soap is best, such as the following: 1 part of 40 per cent. nicotine to 500 parts of water, adding 2–5 lb. of soft soap (according to the hardness of the water) to every 40 gallons of water.

NATURAL ENEMIES. The writer has bred the Braconid *Aphidius* (*Lysiphlebus*) *dissolutus* Nees, nec Haliday (Waterston det.) from *A. fabae* in Britain and the following Hymenoptera are recorded from this aphis in other countries: in France the Braconids *Trioxys heraclei* Hal., *Aphidius cardui* Marsh, *A. fabarum* Marsh; the Chalcid *Asaphes vulgaris* Wlk. and the Cynipid *Alloxysta kiefferi* Pic.; in Italy the Braconids *Aphidius fabarum* Marsh, *A. heraclei* Hal., *Trioxys heraclei* Hal. and the Chalcid *Pachycrepis dubia* Buckt.; in Cyprus the Braconids *Aphidius crisii* Hal., *Trioxys centaureae* Hal. and *Ephedrus validus* Hal.; and in America the Braconid *Lysiphlebus testaceipes* Cress., the Chalcids *Aphelinus semiflavus* How., *Asaphes americana* Gir., *Pachyneuron aphidivorum* Ash. and *P. siphonophorae* Ash. and the Cynipid *Aphidencyrtus* sp. Del Guercio in Italy records the Cecidomyid *Trilobia aphidisuga* Del Guer. as parasitic upon *Aphis fabae*.

Aphis rhamni Boyer

SYNONYMS. *Aphis solanina* Pass.; *A. abbreviata* Patch.

DESCRIPTION. *Apterous viviparous female.* Pale green to pale yellow, colour variable; antennae green or yellow with apices dusky; cornicles cylindrical, short, straight and rather stout, apices darker; cauda rather short and thick with three hairs on each side. In the apterous viviparous female there is a single sensorium on the 5th antennal segment. Proboscis reaches to, or just beyond, the second coxae. Legs with short spine-like hairs on tibiae, tarsi dusky. Length 1–1·2 mm.

Life history. The winter host of this aphis appears to be the buckthorn (*Rhamnus catharticus* and *R. frangulae*), on which the winter eggs are laid. From June onwards, it is very commonly found upon the potato, of which it is occasionally a pest. In Cambridgeshire it occurs in large numbers on the potato plant in company with *Macrosiphum gei* and *Myzus persicae*; these two latter species together with *Myzus pseudosolani* and *A. rhamni* are the chief potato feeding aphides of Great Britain.

Aphis avenae Fabricius

Description. *Alate viviparous female*. Head and thorax brown, abdomen green with black markings, antennae dark, cornicles cylindrical. Length 2·3 mm.

The *apterous viviparous female* is green, antennae green, cornicles deep green to brown. Length 1·2 mm. (Theobald, *Aphididae of Great Britain*, II, 161.)

This aphis sometimes attacks cereals and has been recorded as a pest of oats and barley in Wales and occasionally of wheat in the south of England. According to Theobald, it winters as an egg on the bird cherry (*Prunus padus*) and wild cherry (*P. cerasus*); the eggs hatch in April and the winged migrants fly to cereals and grasses where they live until late summer when they return to the winter hosts. It is possible that some may remain on the grasses throughout the winter. *A. avenae* is parasitised by the Braconid *Aphidius avenae* Hal.

Aphis saliceti Kalt.

Occasionally a pest of willows and osiers, and often does much damage to osier beds. It occurs in June and July; it is unusual in its life history in that the sexual forms occur in the summer time. For full descriptions of the various individuals see Theobald, *Aphididae of Great Britain*, II, 171, and Haviland, *Annals of Applied Biology*, 1920, VI. Two Braconid parasites are recorded from *Aphis saliceti*, these are *Aphidius cardui* Marsh and *A. salicis* Hal.

Genus *Brevicoryne* Van der Goot

Head without prominent frontal tubercles; antennae of six segments; cornicles short, swollen in middle, not much longer than cauda which are broadly conical. (Theobald, *Aphididae of Great Britain*, II, 45.)

Brevicoryne (*Aphis*) *brassicae* Linn. The Cabbage Aphis; Mealy Cabbage Aphis

Though somewhat spasmodic in appearance, this aphis constitutes a serious pest of cabbage and related plants. The young aphides at first are shiny and greenish yellow in colour, after the first moult the colour changes to greenish grey and the white mealy wax which is characteristic of the species is secreted; this covers the body and conceals the true colour. Both oviparous and viviparous females are wingless; the winged generation are without the waxy secretion.

DESCRIPTION. *Apterous viviparous female.* Long-oval, covered with a whitish meal, colour greyish green with eight black spots ranged dorsally along each side of the back, increasing in size towards the posterior; antennae shorter than body, green with black tips; eyes black; legs dark; cornicles black, short, narrowed at apices; cauda small and black. Length 2–2·3 mm.

Winged viviparous female. Body yellowish green, no meal; head and thorax black (Theobald gives the pronotum as brown); antennae as long as body, dark brown; abdomen with a row of dark transverse dorsal marks; legs dusky brown; cauda dark green, hairy; cornicles brown, short; wings short with stout brown veins. Length 2 mm.

LIFE HISTORY. The winter is passed in the egg stage, either in old cabbage stumps, Brussels sprouts, or on Cruciferous weeds, the adult females also occasionally survive the winter. The eggs hatch about April or May, giving rise to wingless viviparous females (the stem mothers) which produce large numbers of living young, each female forming a small compact colony. The winged generations appear about August and the wingless egg-laying females in the autumn and on into the winter.

HOST PLANTS. All Brassicas with a preference for Brussels sprouts and other cultivated Crucifers like turnips and swedes; also many Cruciferous weeds such as shepherd's purse (*Capsella bursa-pastoris*), charlock (*Sinapis arvensis*), field cress (*Isatis tinctoria*), wall brassica (*Diplotaxis tenuifolia*), and *Erysimum* spp.

INJURY TO HOST PLANT. There is no mistaking a cabbage which is badly infested with the mealy cabbage aphis. The insects congregate on both upper and lower leaf surfaces and the whole plant presents a pitiable spectacle. The attack usually begins with isolated colonies on one or two plants. Infested plants can easily be recognised by their wilted and yellowing appearance. As the attack progresses, the leaves become curled and distorted with pale blister-like galls, the plant sometimes being killed.

DISTRIBUTION. Widely distributed throughout the British Isles. The following account of recent occurrences is quoted from reports to the Ministry of Agriculture. Local attacks on cabbage were recorded from most districts in 1925, principally in the south-west, in Worcestershire and in South Wales. In the south-east a serious early attack was checked by heavy rain in August. In 1926 the worst attacks occurred in north Devon, and again in 1927 the aphis was unusually abundant in Devonshire but scarce elsewhere. In 1929 severe attacks by the cabbage aphis were general all over the country, particularly in Bedfordshire and Cambridgeshire.

CONTROL. Destroy all refuse of cabbage plants and keep down Cruciferous weeds as much as possible. Attacks on the seed beds should be dealt with by means of a nicotine soap spray (1 fluid ounce of 98 per cent. nicotine to 10 gallons of water with the addition of 1 lb. of soft soap); this spray can also be applied to the growing crops to within 14 days of picking. A 3 per cent. nicotine dust applied at the rate of 50 lb. per acre between 7 a.m. and 1 p.m. has proved effective. A paraffin emulsion (2 pints of paraffin, 1 lb. of soap to 10 gallons of water) or simple solution of soap and water (6–8 lb. soap to 50 gallons of water) are also recommended. A good combined spray for this aphis, *Aleurodes brassicae* (p. 41) and *Plutella maculipennis* (p. 88), is 10–25 lb. soft yellow laundry soap and 1½–4 lb. lead or calcium arsenate to 100 gallons of water. *B. brassicae* is particularly liable to become a serious pest during a drought when the plants are not in a vigorous state. As a general rule the worst attacks develop in the autumn, the summer attacks sometimes dying out. Towards the end of the season the colonies of aphides become very heavily parasitised, in some cases 80–90 per cent. may be affected.

NATURAL ENEMIES. Like most aphides, the cabbage aphis is preyed upon by the larvae of Syrphids; the following species are recorded as commonly attacking *B. brassicae*: *Syrphus ribesii*, *S. grossulariae*, *S. balteatus* and *Catabomba pyrastris*, and in addition the Coccinellid beetles *Adalia bipunctata* and *Coccinella*

7-punctata. The writer has found the Braconid *Aphidius brassicae* Hal. parasitising large numbers of *B. brassicae* and Buckton records Hymenoptera of the genera *Ceraphron*, *Trionyx* and *Coruna*; a species of *Allotria* (Cynipidae) has also been recorded in Britain. Leonardi (1922, 1925 *Insetti Dannosi e Loro Parassiti in Italia Fino al* 1911) gives the following list of natural enemies of *Brevicoryne brassicae* in Italy:

HYMENOPTERA	DIPTERA
Allotria brassicae Ashm.	*Lasiophthicus pyrastri* L.
Aphelinus mali Hal.	*Syrphus auricollis* Meig.
Lipolexis picea (Cress.) Ashm.	*S. balteatus* De G.
Pachyneuron aphidivorum Ashm.	*S. corollae* Fab.
Aphidius avenae Hal.	*S. fasciatus* Fab.
A. brassicae Marsh.	*S. latifasciatus* Mecq.
A. rapae (Curt.) Marsh.	*S. ribesii* L.
Encyrtus aphidiphagus Ashm.	
Isocratus vulgaris Wlk.	COLEOPTERA
Limnerium rivale (Cress.) D.T.	*Coccinella* spp.
L. tibiator (Cress.) D.T.	*Adonia variegata* v. *carpini* Geoff.
	Adalia bipunctata L.

In North America, the Braconid *Praon simulans* Prov. and the Chalcid *Aphelinus semiflavus* are recorded from this aphis; while in Hawaii the Braconid *Diacretus chenopodiaphidis* Ashm. is recorded.

Two aphides of the genus *Anuraphis* are mentioned briefly. They are *Anuraphis dauci* Fab. and *A. tulipae* Boyer.

Anuraphis dauci

A green species with the head, cornicles and cauda black; antennae and legs yellowish, cornicles short, body often with a whitish meal; there are four small papillae on each side of the body.

The writer has found this species on the tops of carrots in Cheshire, often associated with attacks of the larvae of *P. rosae*, the carrot fly. A severe attack by an aphis, thought to be *A. dauci*, on carrots and parsnips was recorded in Shropshire in 1926, where the crops were grown on a field scale. Much of the crop was entirely destroyed, the remainder being saved by spraying with nicotine and soap.

Anuraphis tulipae

The apterous viviparous female is pale fawn to almost white.

This characteristic aphis is sometimes a pest of carrots, living upon the root underground, which it often splits. *A. dauci* may be an aerial form of this species. (Theobald, *Aphididae of Great Britain*, II, 238.)

One species from each of the genera *Pemphigus* and *Phorodon* are to be noticed.

Pemphigus bursarius Linn. Lettuce Root Aphis; Poplar Gall Aphis

The winged viviparous female is greyish green or pale green covered with a white flocculent meal, the cornicles are not visible and the antennae are short.

This aphis is often a severe pest of lettuces, the root of which is attacked, causing the plants to wilt and die. The winter is spent either on the roots of lettuce or allied plants or in the egg stage on the poplar where the female forms a gall. A gall is formed by a stem mother and in it she produces her young. In July, winged migrants fly from the gall, which splits open, and deposit their young near lettuce or sow thistles (*Sonchus*). The young are pallid yellowish white to pale yellow and mealy.

The writer found this aphis doing much damage to lettuces in Cheshire and Derbyshire in 1925 when it was very prevalent in parts of the midlands. It is fairly common all over the British Isles; severe attacks have been recorded from Kent, Wales, south and south-west of England and in the south-eastern counties. Theobald gives a good account of the lettuce root aphis in "Aphides attacking vegetables and market garden crops ", *Journal of the Royal Horticultural Society*, L, Part 1, 1925, or in the *Aphididae of Great Britain*, III, 245.

Phorodon humuli Schrank. Hop Damson Aphis; Hop Fly

For a description of this species the reader is referred to Theobald, *Aphididae of Great Britain*, I, 273.

LIFE HISTORY. This aphis is an important enemy of the hop. The winter is spent as a black shiny egg on the sloe, bullace or damson and more rarely on the plum. The eggs hatch in March and April giving rise to females, which produce apterous viviparous females which again give rise to living young. In May and June winged migrants are formed which fly to the hop, but occasionally the aphis will breed on the *Prunus* until August before migrating to the hop. In autumn the return migration to the damson takes place, where the sexual individuals are produced and the winter eggs laid. Theobald states that the winged males are often produced on the hop, whence they fly to the damson to fertilise the oviparous females.

HOST PLANTS. Hops, *Prunus* spp. especially sloe (*Prunus communis*), bullace, damson; also nettle (*Urtica dioica*).

DISTRIBUTION. Common in all hop-growing districts, Kent, Sussex, Hampshire, Worcestershire, Herefordshire, Shropshire, Hertfordshire, etc.

CONTROL. Where the attack on the hop is early, the aphides can be killed with a soap and nicotine spray (4 oz. of 98 per cent. nicotine to 100 gallons soap wash) or pyrethrum wash, but attacks coming late in the season when the hops are in cone are difficult to deal with, as spraying is impossible. The hop plant is subject to a serious virus disease known as mosaic, and in connection with the spread of this disease in the hop gardens, all aphides attacking the hop plant are open to suspicion until the identity of the insect vector or vectors is definitely established. (See Chapter XIV.)

REFERENCES

HEMIPTERA—HOMOPTERA

(1) BÖRNER, C. (1921). Die Brutpflanzen der Kohlblattlaus. *Mitt. biol. Reichsanst. Land- u. Forstw.* XXI.

(2) CAMPBELL, ROY E. (1926). The pea aphis in California. *Journ. Agric. Res.* XXXII, No. 9.

(3) DAVIDSON, J. (1927). On some aphides infesting tulips. *Bull. Entom. Res.* XVIII, No. 1.

(4) DAVIDSON, J. (1925). The bean aphis. *Journ. Min. Agric.* XXXII.

(5) DAVIDSON, J. (1914). The host plants and habits of *Aphis rumicis* Linn. *Ann. App. Biol.* I, No. 2.

(6) DAVIDSON, J. (1921). Biological studies of *Aphis rumicis* L. *Bull. Entom. Res.* XII, No. 1.

(7) DAVIDSON, J. (1925). *A List of British Aphides.* Rothamsted Monogrs. Agric. Sci.

(8) DAVIS, J. J. (1915). The pea aphis with relation to forage crops. *U.S. Dept. Agric. Bull.* No. 276.

(9) EDWARDS, J. (1896). *The Hemiptera-Homoptera of the British Islands.* London.

(10) FENTON, F. A. and HARTZELL, ALBERT (1923). Bionomics and control of the potato leaf-hopper. *Agric. Exp. Station, Iowa State College, Agric. Research Bull.* No. 78.

(11) FLUKE, C. L. (1929). The known predacious and parasitic enemies of the pea aphid in North America. *Agric. Exp. Station, Univ. Wisconsin, Res. Bull.* No. 93.

(12) HERRICK, G. W. (1911). The cabbage aphis. *Journ. Econ. Entom.* IV.

(13) HORSFALL, J. L. (1925). The life history and bionomics of *Aphis rumicis* L. *Univ. Iowa, Studies Nat. Hist.* XI, No. 2.

(14) HOUSER, J. S., GUYTON, T. L. and LOWRY, P. R. (1917). The pink and green aphid of the potato. *Ohio Agric. Exp. Station, Wooster, Bull.* No. 317.

(15) MERCET, R. G. (1921). Notas sobre Afelininos (Hym. Chalc.). *R. Soc. Esp. Hist. Nat.* Madrid. Special volume.

(16) PATCH, E. M. (1915). Pink and green aphides of the potato. *Maine Agric. Exp. Station, Orono, Bull.* No. 242.

(17) PATCH, E. M. (1928). The foxglove aphid on potato and other plants. *Maine Agric. Exp. Station, Orono, Bull.* No. 346.

(18) PERGANDE, THEO. (1904). On some aphides affecting grain and grasses of the United States. *U.S. Dept. Agric. Div. Entom. Bull.* No. 44.

(19) PHILLIPS, W. J. (1916). The English grain aphis. *Journ. Agric. Res.* VII, No. 11.

(20) SAVAGE, C. G. (1917). The cabbage aphis. *Journ. Dept. Agric. S. Australia, Adelaide*, XX, No. 7.

(21) SMEE, C. (1922). British ladybird beetles; their control of aphides. *Fruitgrower, Fruiterer, Florist and Market Gardener*, London, Nos. 1376–1378.

(22) SMITH, KENNETH M. (1929). Insect transmission of potato leaf-roll. *Ann. App. Biol.* XVI, No. 2.

(23) SMITH, L. B. (1919). The life history and biology of the pink and green potato aphid. *Virginia Truck Exp. Station, Norfolk, U.S.A. Bull.* No. 27.

(24) SMITH L. B. (1916). The green pea aphid. *Va. State Ent. and Plant Path. Rept.* 1914–15, X.

(25) THEOBALD, F. V. (1923). The aphides attacking cereals in Britain. *S.E. Agric. Coll. Wye, Kent, Bull.* No. 2.

(26) THEOBALD, F. V. (1922). The aphides attacking the potato. S.E. Agric. Coll. Wye, Kent, Advisory and Research Dept.

(27) THEOBALD, F. V. (1926). *The Plant Lice or Aphididae of Great Britain.* Vols. I, II, III.

(28) THEOBALD, F. V. (1925). Aphides attacking vegetables and market garden crops. *Journ. Royal Hort. Soc.* L, No. 1.

(29) WHEELER, E. W. (1923). Some Braconids parasitic on aphides and their life histories. *Ann. Entom. Soc. America*, XVI, No. 1. Columbus, Ohio.

(30) WEED, A. (1927). Metamorphosis and reproduction in apterous forms of *Myzus persicae* Sulz. as influenced by temperature and humidity. *Journ. Econ. Entom.* XX, No. 1.

CHAPTER VII

LEPIDOPTERA

Order 18. **Lepidoptera.** Butterflies and Moths

Two pairs of membranous wings, the body, wings and legs clothed with flattened scales, which give rise to the colours so often found in members of this order; mandibles usually absent, the mouth-parts consisting of a long sucking tube or 'tongue' formed by the maxillae; antennae usually long and many-jointed; metamorphosis complete.

This group is very large and is represented in the British Isles by more than 2000 species but, with one or two notable exceptions, the group as a whole is not of very great agricultural importance.

Sub-order *Rhopalocera.* Butterflies

Family *Pieridae.* White Butterflies

Eggs mostly tall and spindle-shaped, caterpillars with body cylindrical, usually bare or with a few short hairs; pupae with angular projections, fixed to solid objects by the tail and suspended by a silken girdle round the middle.

Pieris brassicae Linn. The Large White Butterfly

DESCRIPTION. *Adult.* White, with fore wings black at tip; in the male a black spot on the upper edge of the hind wings; female with two black spots and an inner marginal dash on the fore wings. Wing expanse 2½–3 in.

Egg. The eggs are deposited in batches of 5–100 and not singly, they are spindle-shaped, standing on end with 15–17 longitudinal ribs, reticulated transversely, in colour they are dull yellow. (Fig. 12.)

Larva. The newly hatched larva is pale green with a black head but when fully developed the head becomes paler, there is a dorsal line of deep yellow and a yellow line along the spiracles,

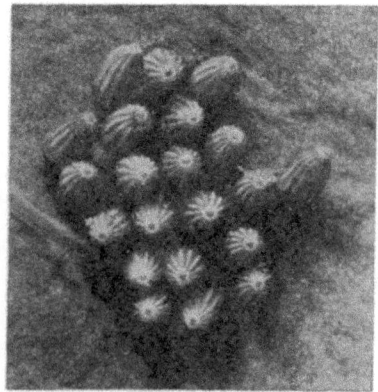

Fig. 12. Eggs of *Pieris brassicae* Linn. The large White Butterfly. Much enlarged.

the under side of the body is greenish, the legs yellow with black markings, the back is bluish green and the body has a row of black spots down each side and is covered with short hairs. Length 24–36 mm.

Pupa. Angulose, grey or greenish in colour with a variable quantity of black spots, there is a pointed projection above the head; fastened with silken girdle and anal pad usually to walls and fences, not suspended.

LIFE HISTORY. The winter is passed in the pupal condition, the adults hatching in April or earlier and occurring throughout the year until October. There are two broods in a season, the second brood being sometimes reinforced by the addition of migrants from the continent. The eggs are deposited in batches of 5–100 and not singly as with the closely allied *P. rapae.* The duration of the egg stage is about 6–8 days; on hatching the young larvae eat the egg shells and live for a time gregariously, later they live singly. The eggs of the first brood are laid in May and those of the second in July and August and sometimes later. The first brood is less numerous and troublesome than the second and it is not so heavily parasitised. There are four moults during the lifetime of the caterpillar which lasts for 4–6 weeks.

INJURY TO HOST PLANT AND SYMPTOMS OF ATTACK. The newly hatched larvae eat off the lower epidermis only of the cabbage leaf, but later holes appear in the leaves, and in a bad attack the entire plant may become skeletonised, only the stalks and veins remaining. As the plant begins to form a heart the caterpillars often go down into it in search of the tender leaves and in these circumstances they are difficult to kill.

CULTIVATED HOST PLANTS. Mostly Crucifers, especially Brassicas such as savoy, sprouts, broccoli, cauliflower, kale, etc.; also turnip, swede, mustard, rape, horseradish, radish, watercress, nasturtium, stock, geranium, and mignonette.

WILD HOST PLANT. Among wild host plants it has been recorded from bitter cress (*Cardamine hirsuta* Linn.).

DISTRIBUTION. Both *P. brassicae* and the allied *P. rapae* are migratory and, though the latter is never absent from this country, *P. brassicae* is sometimes very scarce while at other times it is exceedingly abundant and destructive. It occurs all over the British Isles, throughout Europe and Asia.

Pieris rapae Linn. The Small White Butterfly;
Small Cabbage White

DESCRIPTION. *Adult.* This is the smaller of the two common white butterflies. The adult insect is white, in the male the fore wings are black tipped with a single black spot upon the upper edge. In the female there is an additional black spot on the fore wings, and a black dash on the inner margin of the same. Wing expanse about 2 in.

Egg. The egg is spindle-shaped, pale yellow, with about eleven longitudinal ribs, reticulate transversely.

Larva. When full grown the larva is velvety green with a yellow dorsal line. There is a row of yellow spots along each side in line with the spiracles, which are pale and surrounded with a black ring. The surface of the body is covered with fine black dots bearing pale hairs. There are four moults. Length 24 mm.

Pupa. Angulose, greenish in colour, freckled with black. The pupa is attached by an anal pad and girdle usually to a leaf of the host or near-by object. Length 18 mm.

LIFE HISTORY. There are two broods in the year, the spring brood being the earliest butterflies to appear, hatching sometimes in February and March and regularly in April. The eggs are deposited singly, generally on the under sides of the leaves; the larvae of the first generation pupate in June and July, the second or summer brood, which does most damage, appearing in August. The caterpillars arising from eggs laid by this brood pupate in the autumn and thus pass the winter in sheltered crevices.

CULTIVATED HOST PLANTS. Chiefly Cruciferous plants, especially Brassicas. Experiment has shown the following descending scale of preference: cabbage, radish, nasturtium and mignonette.

WILD HOST PLANT. *P. rapae* has been recorded from *Alyssum* sp.

DISTRIBUTION. Practically worldwide; common all over England, it occurs from the British Isles to Japan, while West has recently recorded *P. rapae* for the first time from New Zealand (*Entom. Mon. Mag.* LXVI, Oct. 1930). Since its first introduction into North America, about 1860, it has spread all over the United States.

CONTROL. Control measures apply equally well to both species of white butterflies. Lead arsenate applied in the form of a dust is an efficient control for the larvae, or as a spray, 1 lb. of Paris green and 2–3 lb. of lime in 100 gallons of water may be used. The use of arsenicals, however, on foodstuffs is open to criticism, and if employed they should be applied only to the seed bed. The alternatives to arsenic are the following: nicotine or nicotine sulphate dust, pyrethrum as a dust (sprays do not adhere easily to cabbage

leaves) or derris powder. Soap and water or salt and water are sometimes useful, the latter should be used at the rate of 2–3 oz. of
common salt per gallon of water, and applied with a knapsack
sprayer. D.D.T., applied as a dust or a spray, might well be
effective against *Pieris* larvae.

NATURAL ENEMIES. There are several Hymenopterous parasites
attacking both species of white butterflies, including Ichneumons
and Braconids. The best known is the Braconid *Apanteles glomeratus* L. This insect, erroneously stated by Fabre to oviposit in the
egg of *P. brassicae* (Smith (27)), oviposits in the body of the caterpillar, the egg soon hatches and the resulting larva lives upon the
fat-body of the host, as many as 150 larvae occurring inside the body
of one caterpillar. When full fed, the parasitic larvae burst through
the skin of the host and construct their bright yellow cocoons
outside the body of the now moribund caterpillar. The attack of
A. glomeratus is always heavier upon the second or autumn brood.

The following list, necessarily incomplete, is given of the parasites of *P. brassicae* and *P. rapae* in Europe.

PIERIS BRASSICAE

In France:

HYMENOPTERA

Ichneumonidae
  Anilastus ebeninus (Gr.) Thoms.
  A. vulgaris Brischke.
Braconidae
  Apanteles glomeratus Linn.
  A. rubripes Hal.
Chalcididae
  Chalcis femorata Panz.
  Dibrachys sp.
  Melittobia creasta Wlk.
  Trichogramma evanescens Westw.

In Italy (Leonardi):

HYMENOPTERA

Ichneumonidae
  Hemiteles cingulator Grav.
  Anilastus ebeninus (Grav.) Thoms.
Braconidae
  Apanteles cajae (Bouche) Marsh.
  A. congestus Nees.
  A. congregatus Say.
  A. glomeratus Linn.
  A. jucundus Marsh.
  A. rubripes Hal.
  A. spurius Wesm.
  Microgaster subcompleta Nees.

Tachinidae   DIPTERA
  Ceromasia flavorum Macq.
  Tachina larvarum
  Tricholyga segregata Rnd.
Drosophilidae
  Drosophila busckii Coq.
  D. rubostriata Beck.

In Germany and Europe generally:

HYMENOPTERA

Ichneumonidae
  Angitia rufipes Grav.
  Pimpla brassicariae Poda.
  P. instigator F.
  Theronia atalanta Poda.
Braconidae
  Apanteles spurius Wesm.
Chalcididae
  Chalcis intermedia Nees.
  Pteromalus puparum L.
  Polynema ovulorum Hal.

Tachinidae   DIPTERA
  Compsilura concinnata Mg.
  Exorista lota Mg.
  Masicera sylvatica Fln.
  Phryxe vulgaris Fall.

In Poland the fungus *Entomophthora
sphaerosperma* is recorded.

PIERIS RAPAE

In Italy (Leonardi):

HYMENOPTERA

Ichneumonidae
Hemiteles melanarius Gr.
Pimpla rufata Gr.
Braconidae
Apanteles congregatus Say.
A. glomeratus Linn.

Braconidae
Microgaster subcompleta Nees.
Chalcididae
Pteromalus brassicae Payk.

DIPTERA

Tachinidae
Compsilura concinnata Mg.
Phryxe vulgaris Fall.

Sub-order *Heterocera*. Moths

Family *Noctuidae*. Owlet Moths

The largest family of the Lepidoptera, it contains a great number of dull coloured moths of medium size; the hind wings are usually paler than the fore wings and the sexual dimorphism is not great. The eggs are rounded with a micropyle at the top, pale in colour, ribbed or sculptured. The larvae are mostly nocturnal feeders and pupate in an earthen cell in the soil; they constitute some of the most serious insect pests of agricultural crops in America. In this country, also, considerable damage is done by the larvae of certain species, known as 'cutworms', which are dealt with herewith.

Euxoa (*Agrotis*) *segetum* Schiff. The Turnip Moth; The Dart Moth

DESCRIPTION. *Adult*. (Fig. 13.) A very variable species. Male. Fore wings brownish grey to reddish brown; claviform mark dark, rather distinct; orbicular mark oblong, very faint; reniform mark usually distinct, outlined in pale brown, ringed with darker; a faint wavy line, parallel to the hind margin, traverses the wing close to the reniform mark, it is lunulate in shape with the concave side towards the hind margin; antennae pectinate; hind wings white with an opalescent shine, nervures and marginal line fuscous.

Female. Antennae filiform; hind wings dirty white shading into dusky at the margin; body somewhat darker. Wing expanse about 30 mm.

Larva. Pale, smoke-coloured or greenish grey, there is a pale dark edged line down the back and faint indications of lateral lines. Four black spots are present dorsally, two on each side of the central lines, while three other spots are present laterally on each segment. The spiracles are small and dusky, *smaller* than the surrounding black spots. Under surface of body light grey. Head greyish with two dark curved lines forming an X. Length 24–36 mm.

Pupa. Smooth, reddish brown in colour, with two spines at the hind end. Length 10–12 mm.

LIFE HISTORY. The winter is passed in the ground in the larval stage, and feeding may persist during the winter in mild weather. The larvae are full fed in May and June when they pupate, the adults emerging in June and July. The eggs are deposited from June onwards, they are laid on the stems of cultivated plants and weeds, close to the surface of the ground. The larvae hatch in 10–14 days and feed for a time on the lower parts of the plant. Later they pass into the ground and feed upon the roots and underground portions of the stem. If the crops are young the larvae emerge at night to feed and gnaw through the stem level with the soil surface, it is this habit which has given rise to the name 'cutworm'. As a rule there is only one generation in the year, but occasionally there may be a partial second brood of moths in late summer, the larvae of this partial brood hibernating in the soil with those of the first brood. The presence of cutworms in a field, therefore, is not due to the crop which is being grown, but to the treatment of the field during the previous year.

INJURY TO HOST PLANT AND SYMPTOMS OF ATTACK. Injury by the cutworm varies according to the nature of the crop; it takes the form of injuries to leaf and stem, hollowing out of potato tubers and succulent underground roots such as turnips and swedes, and the decapitation of seedling crops. The last mentioned may be serious, as the caterpillars are liable to move down the rows destroying the seedlings as they go.

CULTIVATED HOST PLANTS. The larvae of *E. segetum* have a very wide range of food plants, attacking potatoes and root crops such as swedes and turnips, also oats, beans, carrots and all kinds of seedling plants, particularly seedling Brassicas. They have also been recorded as attacking chrysanthemums.

WILD HOST PLANTS. It is probable that this insect can feed upon a very large number of weeds and wild grasses; the writer has found it at the roots of *Lolium* sp.

DISTRIBUTION. Common all over the British Isles, Europe and the United States of America.

CONTROL. Climatic conditions influence the numbers of cutworms to a large extent, and a factor governing the occurrence in epidemic form appears to be the rainfall in late spring, while heavy rains in July seem to be instrumental in checking an attack. A possible explanation of this lies in the fact that cutworms are

74   LEPIDOPTERA

particularly prone to attack by bacterial diseases under conditions of abnormal dampness. It is also possible that repeated cultivation of the soil has brought about a partial change in the habits of cut-worms by enabling them to live beneath the surface, instead of upon the aerial parts of the host plant. This mode of life may place the caterpillars beyond the reach of their natural enemies, and a wet season, by rendering the soil unsuitable, brings them to the surface where they may be attacked by parasites and other pre-dators. This may be one reason why a wet summer is usually followed by a decrease the succeeding year. Light, well-cultivated soils, easily penetrated by the larvae, are, therefore, the most often infested. The method of artificial control practised largely in America is the use of the poison bait which is also applicable to this country. It is the same as that used for controlling Tipulid larvae (p. 161) and consists of the following mixture: 1 lb. of Paris green, 16–20 lb. of bran and ½ pint of treacle or molasses, and sufficient water to make the mixture easy to broadcast. This quantity is sufficient for one acre. An alternative method is to use beet leaves and roots chopped up and dipped in a mixture of Paris green and diluted molasses, placed about the fields in small heaps. A lead arsenate spray (1 lb. of lead arsenate paste to 25 gallons of water) on young root crops is sometimes successful. It is important to destroy weeds before the eggs are deposited, otherwise if weeds bearing the eggs are ploughed under, the eggs will hatch and the resulting larvae prey upon the crop. Poultry are of considerable value in eating cutworms during cultural operations.

NATURAL ENEMIES. The following natural enemies are recorded from *E. segetum* in this country and on the continent. (Herold (16).)

HYMENOPTERA
Ichneumonidae
  Ichneumon bimaculatus Schrk.
  I. sacritorius L.
  Amblyteles armatorius
  A. fuscipennis Wesm.
  A. melanocastanus
  A. panzeri Wesm.
  A. vadatorius Wesm.
  Anomalon cerinops Gr.
Braconidae
  Microplitis seurati Marsh.
  Amicroplus (*Macrocentrus*) *collaris* Spin.
  Apanteles spurius Wesm.
  Bracon kollari Rnd.
    There are also records of three undetermined Braconids.

Chalcididae
  Oophthora (*Pentharthron*) *semblidis* Auriv.
  Trichogramma evanescens Westw.
  T. semblidis

DIPTERA
Tachinidae
  Peletieria nigricornis Meig.
  P. prompta Meig.
  Tachina larvarum L.
  Gonia capitata Deg.
  G. divisa Meig.
  Cnephalia bucephala Meig.
  C. bisetosa B.B.
  Bucentes cristata F.
  Pales pavida Mg.
  Phryxe vulgaris Fall.

75

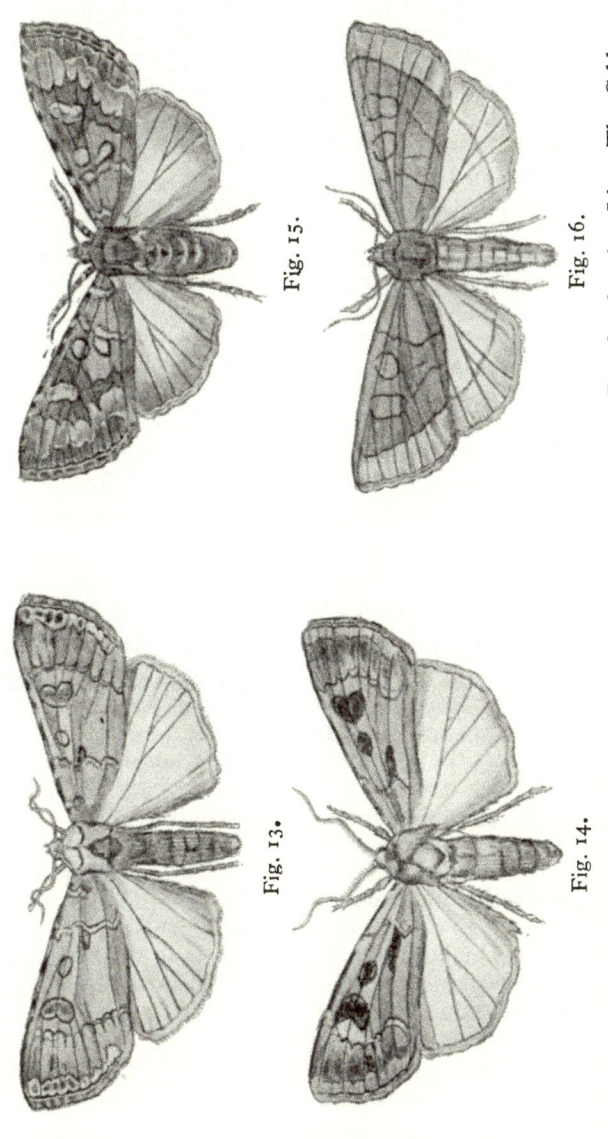

Fig. 15.

Fig. 16.

Fig. 13.

Fig. 14.

Fig. 13. *Euxoa segetum* Schiff. The Turnip Moth. Male.
Fig. 14. *Feltia exclamationis* Linn. Heart and Dart Moth. Male.
Fig. 15. *Barathra brassicae* Linn. The Cabbage Moth. Male.
Fig. 16. *Gortyna micacea* Esp. Rosy Rustic Moth.

All slightly enlarged.

Feltia (*Agrotis*) *exclamationis* Linn. Heart and Dart Moth

DESCRIPTION. *Adult.* (Fig. 14.) Male. Very variable, fore wings reddish, reddish brown or blackish; claviform mark usually large and dark in colour; orbicular mark sometimes well developed, but often absent or in close contact with the reniform mark, which is usually dark and distinct except in the upper part. There are two faint lunulate lines parallel with the hind margin; hind wings white with a faint marginal band.

Female. Hind wings dark, lunule distinct. Wing expanse 24–36 mm.

Egg. Large, rounded above and flattened beneath, thickly ribbed and with fine reticulations, colour whitish at first, changing later to a glistening, dirty flesh colour. (Buckler (4).)

Larva. Very similar to the larva of *E. segetum*, colour variable, dark brown, grey or greenish, ventrally the colour is lighter. Down the back is a row of pear-shaped markings of darker colour with the narrow end towards the posterior. The spiracles are black and *larger* than the spots surrounding them. This character serves to distinguish the larva from that of *E. segetum*. Length 24–36 mm.

NATURAL ENEMIES. The Ichneumon *Amblyteles panzeri* Wesm. is recorded in Italy.

Tryphaena pronuba. Large Yellow Underwing Moth

DESCRIPTION. *Adult.* Very variable, fore wings range from pale brown to dark umber; the stigmata are sometimes definite sometimes lost among numerous wavy lines. The most constant characters of the fore wings are the pale wavy line parallel with the basal margin and a double black spot near the upper extremity. The hind wings are yellow with a narrow black border, there is no central spot which distinguishes *pronuba* from the allied species *orbona*. Wing expanse 30–40 mm.

Egg. Circular, rather flattened above and below, strongly ribbed and reticulate, pinkish at first later turning to grey. (Buckler (4).)

Larva. Very variable in colour, yellowish or greenish grey. There is a light stripe down the back which is darker than the rest of the body, and a light-coloured, pale yellowish or greenish stripe along each side with conspicuous black marks above it; spiracles black or dark yellow rimmed with black. Length 48 mm.

Pupa. Shining, reddish brown, posteriorly there are two spines with two smaller points.

The life histories and methods of control are the same for *F. exclamationis* and *T. pronuba* as for *E. segetum*, except that

T. pronuba appears later in the season, although all three species may be found together.

HOST PLANTS. Theobald (*Journ. S.E. Agric. College*, 1929) records an attack upon hops; the egg masses were laid on the leaves in August and the larvae hatched out in September, hibernating shortly after. The food plants are dandelion, docks, lettuce and various Brassicas.

NATURAL ENEMIES. Fahringer(9) records the following Hymenopterous parasites of *T. pronuba* as occurring in Europe: the Ichneumonids *Amblyteles armatorius* Forst.; *Ichneumon confusorius* Grav., and the Braconid *Apanteles congestus* Nees. Hase records the Chalcid *Trichogramma evanescens* Westw. in Germany, while Leonardi gives the following additional Hymenoptera from Italy: the Ichneumonids *Amblyteles monitorius* Wesm., *A. quadripunctorius* Berth., *A. vadatorius* Wesm., *Ichneumon extensorius* Linn., *I. fusorius* Linn., *I. pisorius* Linn. and the Braconid *Meteorus versicolor* Ruthe.

Barathra (*Mamestra*) *brassicae* Linn. Cabbage Moth

DESCRIPTION. *Adult.* (Fig. 15.) General colour ashy grey, varying to darker or almost black, often with an ochreous tint; the reniform mark is distinct with a white outline and white sub-marginal line; orbicular mark with a black margin; claviform mark often indistinct. There is a white transverse zigzag line across the fore wing near the basal margin; hind wings brown with a central spot. This species is exceedingly variable and many distinct varieties are described under different names. Wing expanse 42 mm.

Larva. When first hatched the colour is green, later turning to various shades of dark green, grey or brown. There are three light lines on the back with oblique black dashes on each segment, on the 12th segment is a black horseshoe-shaped spot. The lateral line is yellowish, there is a sharp differentiation in colour above and below the spiracular line.

LIFE HISTORY. The moths appear in May and June and oviposit on the leaves of cabbages or other host plants, the globular eggs are deposited singly on the leaves. The caterpillars are full grown in 4–5 weeks when they pupate in the soil, the winter thus being passed in the pupal stage. As a rule there is only one brood in the year, but occasionally a partial second generation may develop, the caterpillars of which feed up to October or even later. The moths fly from May to September and the caterpillars occur from June to October.

HOST PLANTS. The larvae are almost omnivorous, feeding on low-growing plants of all kinds. Besides the cabbage tribe, they have been recorded as pests of turnips, lettuces, rape, peas, onions and green tomatoes. The larva also feeds on a large number of weeds, especially the dock (*Rumex*).

INJURY TO HOST PLANT AND SYMPTOMS OF ATTACK. The injury caused is similar to that due to the larvae of the cabbage butterflies, though the caterpillars of the cabbage moth bore into the heart of the plant to a greater extent and are more injurious on that account as they foul the tender leaves with their excrement and render them unfit for food.

DISTRIBUTION. Occurs all over the British Isles and Europe generally, it is a serious pest in Russia.

CONTROL. The methods of control recommended for the larvae of the cabbage butterflies are equally applicable to the control of this insect. Also re-ploughing affected fields in autumn or early spring to expose the wintering pupae is a useful measure; hand picking and the use of poultry will help to reduce the numbers.

NATURAL ENEMIES. Fahringer records the following Ichneumonids as parasitising *B. brassicae* in Europe, *Amblyteles armatorius* Forst., *A. palliatorius* Grav., *Exetastes fornicator* F., *Ophion luteus* Linn., *Paniscus testaceus* Grav. The Braconid *Microplitis tuberculifera* Wesm. and the Chalcid *Cratotechus larvarum* Linn. are recorded from France, while the Chalcid *Trichogramma evanescens* Westw. occurs in France, Germany, and Great Britain. Leonardi records the following list of Hymenoptera from *B. brassicae* in Italy: the Ichneumonids *Amblyteles armatorius* Holm., *Phygadeuon fumator* Grav., *Pimpla brassicariae* Pogh., *P. examinator* Grav., *P. instigator* Grav.; the Braconids, *Apanteles congestus* Reinh., *A. glomeratus* Reinh. and the Chalcid *Pteromalus puparum* Swed.

Baer(1) records the following Tachinids (Diptera) parasitising *B. brassicae* in Europe: *Bucentes geniculata* De G., *Compsilura concinnata* Meig., *Phorocera assimilis* Fall., *Tachina larvarum* Linn., *Voria ruralis* Fall.

Hadena (*Apamea*) *basilinea* W.V. Rustic Shoulder Knot

DESCRIPTION. *Adult.* Ground colour pale whitish ochreous to reddish brown. There is a characteristic black streak in the centre of the base of the fore wing (apparently the 'shoulder knot' of the popular name). Stigmata pale; the reniform mark margined with white with two white dots below. Hind wings brown with a slightly darker margin. Wing expanse 30–36 mm.

Larva. Green to olive brown; head light brown, shining; there are two brownish lines down the centre of the back which is whitish; the second and last segments bear a brown plate on the dorsal surface; laterally two black tuberculate dots are placed obliquely on segments 3–12; under side greenish. The larva tapers somewhat at each end. Length 24 mm.

LIFE HISTORY. The eggs are laid in June, often in the ears of wheat or on grasses; the larvae migrate from the grasses and ascend the wheat stems to feed on the grain, which is eaten in both milky and hardened condition. The larvae hibernate, often in the soil; they recommence feeding in the spring and attack the leaves and stem of the wheat plant boring into the stem and eating it through. Curtis observes of this species that three caterpillars were capable of destroying 30 stems of wheat in 14 days.

INJURY TO HOST PLANT AND SYMPTOMS OF ATTACK. The injury takes two forms, the hollowing out of the wheat grains in late summer and the destruction of the stem in spring. The first type of injury is recognised from the holes evident in the grain and spikelets, while, in the second type, the presence of the caterpillar in the stem is indicated by the widening of the sheath in the last mature leaf, and the yellowing of the youngest protruding leaf due to its being cut through by the larva.

DISTRIBUTION. Very common and widely distributed over the United Kingdom, Northern and Central Europe. Severe attacks on cereals are not frequent but sporadic outbreaks indicate that *H. basilinea* is a potential pest of cereals.

HOST PLANTS. Cereals, especially wheat and barley; it also attacks many wild grasses.

NATURAL ENEMIES. The Braconid *Microplitis tuberculifera* Wesm. is recorded as a parasite of *H. basilinea* in Europe.

Polia (Hadena) oleracea Linn. Bright-line Brown-eye

DESCRIPTION. *Adult.* Not a very variable species; general ground colour of fore wings reddish or fuscous brown; there is a well-defined white line with a **W**-shaped mark parallel with the basal margin of the fore wings; reniform mark yellow or ochreous;

orbicular usually rimmed with white; hind wings brown, darker in the female.

Larva. Greenish or brownish spotted with white, there are three indistinct lines on the back, the outer ones pale; the lateral line is yellow with a blackish margin on which lie the spiracles, which are white with a black rim.

LIFE HISTORY. The winter is passed in the pupal state in the soil, the moths hatching in June and July. The eggs are deposited mostly on the under sides of leaves and the larvae are found feeding in August and September. Pupation takes place in the soil, at a depth of about 1 in.

HOST PLANTS. Cabbage, nettle (*Urtica dioica*), dock (*Rumex* sp.), mignonette (*Reseda* sp.), swedes.

NATURAL ENEMIES. Leonardi records the following parasites from *P. oleracea* in Italy:

HYMENOPTERA	Braconidae
Ichneumonidae	*Apanteles spurius* Reinh.
Amblyteles castigator Wesm.	*Blacus humilis* Nees.
Exetastes cinctipes Thoms.	*Meteorus deceptor* Ruth.
E. fornicator Grav.	*Microplitis spinolae* Reinh.
E. nigripes Grav.	DIPTERA
Ichneumon deliratorius L.	*Compsilura concinnata* Meig.
Braconidae	*Pelatachina tibialis* Fall.
Apanteles ruficrus Hal.	*Tricholyga grandis* Zett.

Chevalier records the Chalcid, *Cratotechus larvarum* Linn. as a larval parasite of *P. oleracea* in France.

Phytometra (Plusia) gamma Linn. Silver Y Moth

DESCRIPTION. *Adult.* Fore wings of marbled appearance, ground colour silvery grey to reddish grey with a velvety sheen. The Y mark is distinct and silvery in colour; hind wings brownish with a darker border. Wing expanse 36–40 mm.

Larva. Varying shades of green, there is a dark green dorsal line with a paler line of whitish green on each side; spiracular line yellowish edged above with green. Some forms have a number of white spots. Head with black markings.

LIFE HISTORY. The caterpillar feeds from June to October on vegetables and other low-growing plants. On account of this exposed method of feeding, dusting the plants with a stomach poison might be an effective control.

NATURAL ENEMIES. The following Hymenopterous parasites are recorded from *P. gamma* in Europe: the Braconids, *Apanteles congestus* Nees., *A. ruficrus* Hal., the Ichneumonids *Pimpla exami-*

nator F., *Exetastes gracilicornis* Gr. and *Opius nitidulator* Nees.
and the Chalcid *Trichogramma evanescens* Westw., the two last-
named from Germany. To these may be added the following
Ichneumonids recorded by Leonardi(17) from Italy: *Amblyteles
sputator* Wesm., *A. ordinarius* Ratz., *A. pallidipes* Reinh., *Ichneumon
comitator* Linn., *I. nigritarius* Grav., and *Pimpla brassicariae* Rogh.
Baer records the following Tachinids from *P. gamma* in Europe:
Bucentes cristata F., *Compsilura concinnata* Meig., *Lydella nigripes*
Fall., *Nemorilla maculosa, Pales pavida* Meig., *P. pumicata* Meig.,
Petina erinaceus F., *Phryxe vulgaris* Fall., *Plagia ruralis* Fln.,
Voria ambigua Fall., *V. ruralis, Tachina larvarum* Linn.

Gortyna (*Hydroecia*) *micacea* Esp. Rosy Rustic Moth.
Potato Stem Borer

DESCRIPTION. *Adult.* Ground colour of fore wings varies from
pink to reddish brown, the stigmata are of the same ground colour
but are ringed with brown. There are two narrow lines parallel
with the basal margin, there is a broad margin of grey for about one
quarter of the fore wing, beginning at the outer margin and terminated by the rosy colour of the rest of the fore wing. Hind wings
grey with a marginal line, a second line traverses the centre of the
wing and appears as a continuation, when the wings are opened,
of the grey margin of the fore wings. (Fig. 16.) Wing expanse
24–36 mm.
　　Larva. Body flesh coloured or pinkish with a reddish stripe on
the back, the segments bear on the sides a number of dark brown
wart-like spots, each with a bristle; the head is reddish brown. On
the second segment there is a polished, brown, semicircular plate
and a small ochreous plate on the anal segment. The spiracles are
black, the under surface of the body pale. Length 24–36 mm.
　　Pupa. Rather stout, ending in an anal spike and a few fine, backwardly directed bristles on the last two segments.

LIFE HISTORY. The eggs are deposited on the lower leaves of the
host plant late in the summer, the winter being probably spent in
the larval state in the soil. The moths hatch out in August and
September, occasionally as late as October. The larva tunnels in
the stem of the host plant, working with great rapidity and completely hollowing it out. When full grown the larva leaves the
plant and pupates in an earthen cell in the soil; this takes place in
July and pupation usually lasts about five weeks.

CULTIVATED HOST PLANTS. Potatoes, green tomatoes, hops, and,
on the continent, strawberries and rape.

WILD HOST PLANTS. Mainly sedges and marsh plants, couch grass (*Agropyron repens*), plantains (*Plantago* spp.), horsetails (*Equisetum* spp.), docks (*Rumex* spp.).

INJURY TO HOST PLANTS AND SYMPTOMS OF ATTACK. In the case of potato, the larva enters the stem from below and works its way up, hollowing it out completely, so that the plant wilts and dies. Theobald (32) describes an attack upon the hop: in the case of this host the young larvae first feed on the tufts of young leaves, spinning them together, they then eat a round hole in the shoot near the apex, and tunnel downwards killing the tips.

CONTROL. This insect occurs sporadically as a pest; the writer has found it damaging large numbers of potatoes in Lancashire. The only method of control appears to be the collection and burning of infested plants.

NATURAL ENEMIES. The Braconid *Microgaster alvearius* F. is recorded from this insect.

Charaeas graminis Linn. The Antler Moth

DESCRIPTION. *Adult.* (Fig. 17.) Variable, dull grey to reddish brown; reniform stigma white and three-branched somewhat like

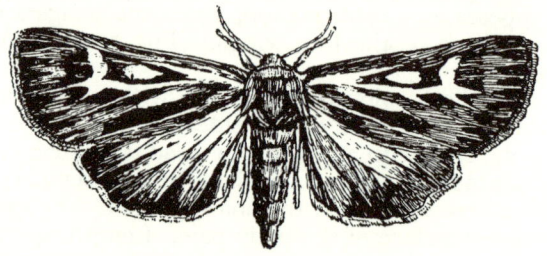

Fig. 17. *Charaeas graminis* Linn. The Antler Moth. × 2.

the antlers of a stag; central line white; orbicular and claviform white. There are often darker markings on the wings but these may be absent; hind wings brown; antennae pectinate in male. Female generally larger; antennae filiform. Many varieties exist, which are differentiated mainly by variations in colour and the stigmata. Wing expanse 24–36 mm.

Larva. Bronzy brown in colour, glossy and much wrinkled. Buckler (4) describes it thus: "a pale line running between the spiracles and the subdorsal stripe. In this species the segmental folds offer a good diagnostic character, being smoother and of a different tint from the back, in fact catching the eye as narrow transverse bands, the whole skin also is much wrinkled transversely and there are transverse pale streaks in the space alluded to between the subdorsal and subspiracular stripes, i.e. three above the pale line and two below it on each segment. The subspiracular stripe is wider than in allied species and the belly seems to have rather a pale golden brown gloss".

LIFE HISTORY. The eggs are deposited in the grass during August by the female moth while in flight. Some doubt appears to exist as to the exact details of the life history. It has been stated by one authority that the winter is passed in the egg stage, and by another (Ormerod) that the partially fed larvae hibernate as such and resume feeding in the spring. It is probable that hibernation does usually occur in the larval state. Pupation takes place in June, the adult moths appearing in July and August. The caterpillars of the antler moth belong to the so-called 'army worms' or 'processionary caterpillars' which appear occasionally in vast numbers migrating from field to field devouring the crops as they go. C. graminis feeds upon pasture and usually appears in mountainous districts.

HOST PLANTS. All kinds of grasses, of which the larvae eat away the shoots and roots; they occasionally attack cereals such as oats, barley and rye. According to Service the plants most affected in Scotland are the tufted scirpus or deer's hair (Scirpus caespitosus Linn.), the rushes Juncus articulatus Linn. and J. squarrosus Linn., bent grass (Agrostis vulgaris Linn.), the purple molinia (Molinia caerulea Moench.), wire bent grass or mat grass (Nardus stricta Linn.). It would appear that the meadow foxtail (Alopecurus pratensis Linn.) and purple clover (Trifolium pratense Linn.) are avoided by the larvae.

DISTRIBUTION. Epidemics of this insect appear occasionally, a bad attack occurred in 1917 in the Peak district of Derbyshire and was investigated by Cole and Imms(6). In 1927 an outbreak was recorded on mountain pastures on the Plynlimmon Range at an elevation of 2000 ft. This attack was largely controlled by seagulls.

CONTROL. Control measures, when necessary, should take the form of trenches with vertical sides cut in the path of the advancing larvae, also poison baits as used for cutworms might prove efficacious. If the attack is on such a scale that several square miles are involved, then firing the grass appears to be the only remedy.

NATURAL ENEMIES. The following Ichneumonids are recorded from *C. graminis* in Italy (Leonardi): *Ichneumon bucculentus* Wesm., *I. gradarius* Wesm., *I. impressor* Zett., *I. molittorius* Linn. and *Pimpla arctica* Zett., and the Ichneumonid *Coelichneumon impressor* from Sweden.

Family *Hepialidae*. Swift Moths

Medium sized moths, head and thorax with woolly hair, abdomen long, extending beyond hind wings; antennae very short; larvae slender, devoid of pattern, with shining head, mostly subterranean or wood feeders.

Hepialus humuli Linn. Ghost Moth; Large Swift; Otter Moth

DESCRIPTION. *Adult. Male.* An insect of characteristic appearance, the wings narrow and rather pointed, white above with a silky sheen, under-surface brown. Wing expanse 48 mm.

Female. Similar in shape; fore wings dull yellow or brownish, with two oblique brick red interrupted stripes; hind wings dull reddish or smoky. Wing expanse 48–60 mm.

Egg. Small, broadly oval, shell smooth, shining white at first, later turning black. Length 0·7 mm.

Larva. Head oval, reddish brown, shining; body creamy white in colour with numbers of tuberculate spots each with a short hair; there is a darker line down the back as far as the 10th segment; on the 2nd segment is a large brown plate; spiracles black; thoracic legs rather long and thin. Length 36–48 mm.

Pupa. Brown in colour and of characteristic shape, stout and blunt-ended, possessing two rows of spines. In general appearance it recalls the pupa of *Tipula* spp. (p. 159.)

LIFE HISTORY. The moths appear in June and July and fly at dusk. The eggs are dropped singly by the female while in flight; the young larvae hatch in 10–14 days and feed on the rootlets of the plant, later they tunnel in the larger roots. Feeding goes on through the winter until April and May, when the larvae pupate in the soil, the adults emerging about a month later. This is an exceedingly common species all over the British Isles and Central Europe. Large numbers of the larvae in different stages may be dug up during the winter at the roots of grass, docks, dandelions, dead-nettle, horehound, etc. It is sometimes a severe pest of the hop. Control consists in clearing away patches of nettle, burdock or neglected greens in the vicinity of hops, and in hand picking of the larvae from the soil; this can be easily done owing to the large size and characteristic appearance of the larvae.

Hepialus lupulinus Linn. Small Swift; Garden Swift

DESCRIPTION. *Adult.* Male. Fore wings dull brown, fulvous; there is an irregular white streak running from the tip to the inner margin and thence to the base of the wing, and also a long white spot in the centre of the fore wings. These white marks are sometimes absent, hind wings smoke-coloured with a lighter fringe. The female is larger, paler and the markings are less distinct. Wing expanse 24–36 mm.

Larva. When full grown, the larva is greenish to yellowish white in colour with four black warts bearing a bristle dorsally, and four more laterally on each segment; the spiracles are black. The head is brown and there is a horny plate on the 2nd–4th segments. The intestine is often visible dorsally as a brown line. This larva is exceedingly active, crawling backwards as easily and rapidly as forwards. Length 17–20 mm.

Pupa. Pale reddish brown, cylindrical, antennal cases prominent, abdominal segments clearly defined and very mobile, there are two transverse ridges on five of the abdominal segments, tail somewhat truncate with two tubercles. Similar to the pupa of *H. humuli*, but considerably smaller. Length 17–18 mm.

LIFE HISTORY. The habits of this species are very similar to those of *H. humuli*, the moths usually appearing a little earlier. The larva also feeds upon the same weed hosts, but is often a pest of potatoes, parsnips and carrots. The writer has found it on several occasions damaging the latter crop by hollowing out the root.

Family *Pyralidae*

Very small slender moths with long fore wings and legs; the approximation or fusion of the veins Sc and R_1 with Rs in the hind wings readily distinguishes the Pyralidae. (Imms, *Textbook of Entomology*.)

Phlyctaenia (*Pionea*) *forficalis* Linn. The 'Garden Pebble'

DESCRIPTION. *Adult.* Fore wings pale yellow-brown with two almost parallel transverse lines of darker brown which thicken near the apex of the wing forming two brown marks. Hind wings paler with a thin brown marginal line, while an inner indeterminate line runs parallel to this. (Fig. 18.) Wing expanse 25 mm.

Larva. Yellowish green, dorsal line darker, spiracular line dark green; head yellowish.

LIFE HISTORY. The bionomics of this insect have not been studied in the British Isles, and the following facts are derived from a study of *P. forficalis* by Zorin (38) in Russia, and are probably partly applicable to conditions in England. In Russia there is usually only one generation a year though in exceptionally hot summers there may be a partial second brood, the larvae occurring from July till autumn. The winter is passed in the larval state in the pupal chamber; pupation occurs in June and the adults emerge about the middle of the month. In England the moths are found from May

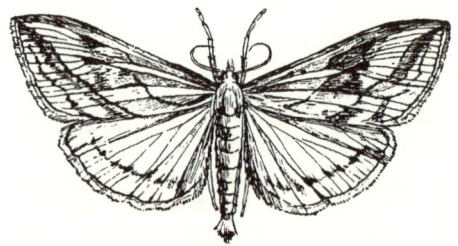

Fig. 18. *Phlyctaenia forficalis* Linn. The 'Garden Pebble'. × 2.

to mid-September, so that there appear to be two broods in this country, the larvae occurring about midsummer and again in the autumn. The eggs are deposited in batches usually on the lower surface of the leaves, chiefly of Cruciferous plants; the larvae also feed on the under surface. The edges of fields appear to be the most generally affected.

DISTRIBUTION. Exceedingly common in gardens and widely distributed in the United Kingdom. It appears to be increasing as a pest of Cruciferae and outbreaks were recorded in different parts of the country in the summer and autumn of 1925–6.

HOST PLANTS. Many Cruciferous plants, especially such Brassicas as cabbage, broccoli and cauliflower; it also occurs commonly on horseradish (*Cochlearia armoracia* Linn.).

CONTROL. As the larvae feed for the most part on the lower surface of the leaf, dusting is not likely to prove efficacious; in the case of a severe attack, spraying with a stomach poison or a strong solution of soap and water might be worth while. The adult moths are attracted to light traps.

NATURAL ENEMIES. The following Hymenopterous parasites have been recorded from *P. forficalis* in Europe: France, *Apanteles astraches* Marsh, *A. gabrielis* Gaut and Riel., *Trichogramma evanescens* Westw.; Italy, *Apanteles picipes* Marsh (also recorded from Great Britain), *A. spurius* .Wesm., *Limnerium geniculatum* Grav., *Ophion minutus* Krchb., *Theronia atalanta* Krieg.

Family *Tortricidae*

Very small moths with wide wings, which are square ended. When at rest the wings fold above the body in a roof-like manner. The larvae live concealed, many in rolled-up leaves.

Cydia nigricana Steph. The Pea Moth

DESCRIPTION. *Adult.* Fore wings unicolorous, brown, satiny, a black line runs close to, and parallel with, the basal margin, while along the basal part of the anterior margin are some characteristic black and white markings; hind wings brown with a pale fringe. Wing expanse 12–15 mm.

Larva. Pale, yellowish in colour, head black, a brown ring on the segment next to the head and eight brown dots on the following segments; legs black. (Ormerod.) Length 6 mm.

LIFE HISTORY. The moths are on the wing from June to August, the majority appearing in the first weeks of July. The eggs are deposited singly on the pea plant, mostly on the upper surface of the sepals of partly developed pods, more rarely on the stem and leaves. Hatching takes place in about eight days (Brittain (3) gives the period as 2–3 days in the laboratory) and the larvae bore through the pod into the developing peas where they feed for about 17–20 days, not more than two larvae being found in one pod. When fully grown and before the peas are ripe, the larvae pierce the side of the pod and descend to the ground, where they form an oval silken cocoon. In this the larva hibernates until the following May or June when it pupates, the adult insect emerging in June thus completing the life cycle.

DISTRIBUTION AND CONTROL. The pea moth is a serious pest locally in Cambridgeshire and the pea-growing districts of the eastern midlands. It was exceptionally troublesome in 1925 in the eastern, south-eastern and southern provinces, late mid-season peas often having 50 per cent. of their crop damaged. In Cambridgeshire it has been found that 'Gradus' peas sown on the 8th and 22nd of April showed only half the percentage attack as compared with those sown on the 6th of May. In Lancashire also, the earliest consignments were practically free from attack while later ones were unfit for market. Cultural remedies should take the form of early sowing, and deep autumn ploughing to destroy the hibernating larvae; also discing the soil after harvest to give birds access to the larvae is recommended. Avoid planting peas two years in succession on the same ground. As regards insecticides, experiments made to control the pest with sprays and dusts have not been very successful, owing partly to the difficulty of making the chemical adhere to the slippery pods. Calcium arsenate has given some measure of control. The following spray, to be applied in the third week of July, is recommended by Miles[21]: calcium caseinate 2 lb., lead arsenate 2 lb., nicotine 10 oz., water 100 gallons.

Family *Tineidae*

A very large group, the majority of which are small moths seldom exceeding a wing expanse of 24 mm. The wings are long and narrow with very long fringes, antennae simple.

Plutella maculipennis Curtis (*cruciferarum* Z.). The Diamond-back Moth; The Small Cabbage Moth

DESCRIPTION. *Adult.* First described by Curtis in 1831. It is a very small insect, the fore wings are light brown with three yellowish triangular marks along the posterior margin of each wing; when the wings are at rest these marks coincide, forming the characteristic diamond-shaped pattern down the back from which the popular name is derived. In this position the wings cover the body and turn up at the end forming a tuft. The hind wings are grey and narrow with long fringes. Female lighter in colour. Wing expanse 16 mm. (Fig. 19 A.)

Egg. Very small, white or greenish yellow, the shell finely reticulate.

Larva. When first hatched the larva is grey with a black head, but in the mature larva the head is slightly chitinised and of a light ochre colour with irregular darker markings on the epicranium, but these latter are not always present. The body is light green, somewhat spindle-shaped. The first segment next to the head has two black spots and the two following segments have a pale spot on each side of the middle line, there are four pairs of slender abdominal legs, the last pair directed backwards; the first

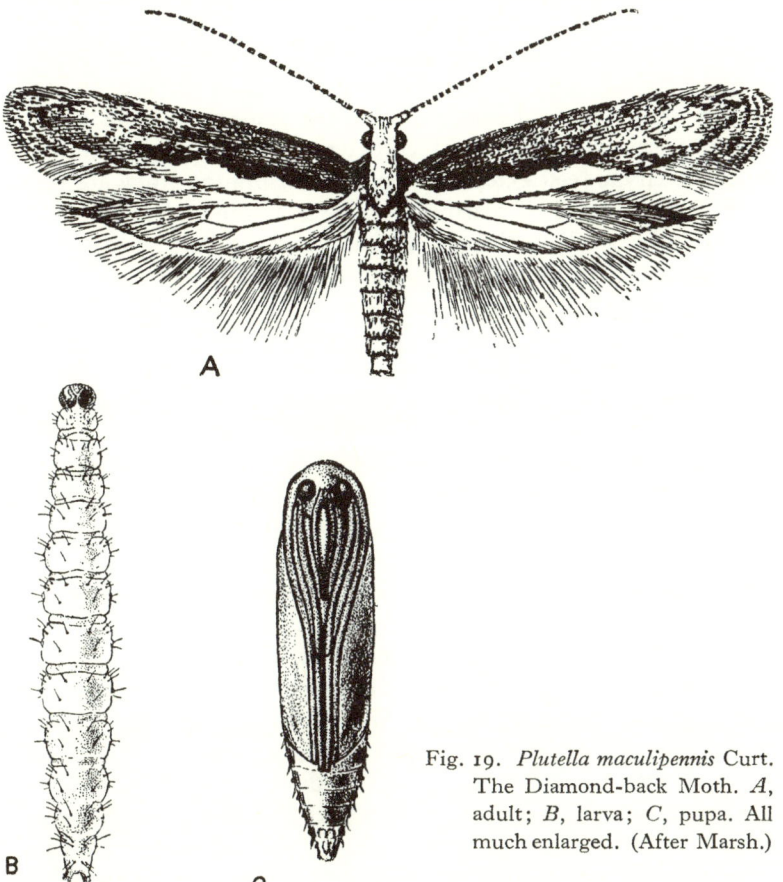

A

B

C

Fig. 19. *Plutella maculipennis* Curt. The Diamond-back Moth. *A*, adult; *B*, larva; *C*, pupa. All much enlarged. (After Marsh.)

pair of thoracic legs is shorter than the second and considerably shorter than the third; there are numbers of small bristles scattered about the body. Length 12 mm. (Fig. 19 B.)

Pupa. Yellowish white or green in colour, sometimes with brown markings, enclosed in a flimsy cocoon open at both ends. (Fig. 19 C.)

LIFE HISTORY. It is fairly certain that in the British Isles the winter is passed in the pupal condition, although in South Africa some writers consider that hibernation takes place in the adult stage. The moths appear in May and June, the first females ovipositing on Cruciferous weeds, later on cultivated Crucifers. The eggs are deposited mostly on the under sides of the leaves in the angles of the veins, either singly or in groups of 2–5; each female is capable of laying 70–90 eggs. The eggs hatch in about 6–10 days and the young larvae at first tunnel in the epidermis of the leaves; later they leave the mines and feed externally. On thin-leaved host plants the larvae feed externally from the beginning; they are very active creatures and, if alarmed, wriggle off the leaf and remain suspended by a silken thread. The larval stage lasts 16–21 days, and there are three moults. Pupation takes place in a fragile silken cocoon, open at both ends, usually upon the leaves of the host or upon dead leaves, etc., near by. The duration of the pupal period during the summer is about 11 days, the total life cycle from egg to adult being about 34 days. As a rule there are two generations a year with occasionally a partial third. This insect is very susceptible to moisture and is only abundant in hot and dry seasons, the numbers are greatly affected by weather conditions, rain having a very deleterious effect upon the larvae.

CULTIVATED HOST PLANTS. All the cultivated Cruciferae, especially swedes and turnips. In Russia it is a pest of beet and it has also been recorded from wallflowers and stocks.

WILD HOST PLANTS. Many Cruciferous weeds, especially charlock (*Sinapis arvensis*) and hedge mustard (*Sisymbrium officinale* Scop.). It also lives upon the prickly saltwort (*Salsola kali* Linn.).

SYMPTOMS OF ATTACK AND INJURY TO HOST PLANT. The larvae devour the leaves of the plant, often reducing them to skeletons, they also attack the hearts of cauliflowers and feed upon the seed heads of swedes, turnips and cabbages. These larvae are also thought to be responsible for a certain amount of 'blindness' in young cabbages. The most serious attacks are on swedes and turnips and develop in July, especially after a spell of hot dry weather.

DISTRIBUTION. Common all over Britain as far as the Shetlands, abundant in Ireland and in Europe generally, especially Russia; it is also present in Asia, Africa, America, Australia and New Zealand. This wide distribution is probably partly due to artificial reasons, the insect being able to adapt itself to many different climates. In this country the worst attacks are always near the sea coast, especially the south-east and east coasts. Fryer (*Insect Pests of Crops*, 1925–7, Min. of Agric.) says of this insect: "the ecology is interesting in that coastal conditions suit it best, not only on the south-east but west and north-west so that migration from the European continent cannot be the explanation of this abundance on the sea coast".

CONTROL. *P. maculipennis* is largely affected by climatic conditions, its development being retarded by wet and cold weather, a heavy rainstorm or a few days of low temperature being sufficient to prevent an attack developing. The following control measures are suggested: soot and lime, three parts soot to one part lime, dusted over the crop at the rate of 2–6 bushels per acre; brushing the leaves by means of branches attached to a scuffler, a horse hoe should follow to bury the dislodged larvae. Theobald recommends one of the following sprays: lead arsenate, this is most suited to the seed bed; nicotine and soft soap; paraffin emulsion. Dusting is also sometimes recommended, and the following might be tried: 1 lb. calcium arsenate to 5 lb. of a 2 per cent. nicotine sulphate dust, this should be applied to the seed bed if possible; soap solution against the young larvae has proved effective on occasions. Austin (*Journ. S.E. Agric. Coll.* 1929) reports some success with an alcoholic extract of pyrethrum as a spray. As with all insects attacking cultivated Crucifers, it is important to keep down so far as practicable any neighbouring wild Cruciferous weeds which might act as alternate hosts or sources of infection.

NATURAL ENEMIES. The most important Hymenopterous parasites are the Ichneumonids *Limnerium gracilis* Gravenh. and *Angitia fenestralis* Holmgr.; the latter may parasitise from 10–90 per cent. In Germany the following parasites are recorded (Torka(34)): *Angitia chrysosticta* Gmel., *A. fenestralis* Holmgr. (the second of these may be only a colour variety of the first), *Diadromus subtilicornis* Grav., *Thyracella collaris* Grav., *T. collaris* var. *brischkei* Berth., *Exochus erythronotus* Grav.; in Italy, *Herpestomus plutellae* Ashm. and *Limnerium tibiator* Cress.; in Sweden, *Diadromus varicolor* var. *intermedius* Wesm.; and in Russia, *Angitia armillata* Grav., *A. fenestralis* Holmgr., *A. gracilis* Grav., *A. majalis* Grav., *Apanteles plutellae* Kurd., *Limnerium blackburni*, *L. polynesiale* Cam., *L. tibiator* Cress., *Phygadeuon rusticellae* Bridg., *Sagaritis latrator* Grav., *Tamelucha plutellae* Ashm., *Tetrastichus sokolowskii* Kurd. Fahringer records the Braconid *Apanteles circumscriptus* Nees. from Europe, while Marsh (18) in the United States of America records *Angitia plutellae* Vier., *Meteorus* sp., *Mesochorus* sp. and *Microplitis* sp. Birds, especially starlings, sparrows and green plover feed upon the larvae.

REFERENCES

LEPIDOPTERA

(1) BAER, W. (1920). Die Tachinen als Schmarotzer der Schädlichen Insekten. Ihre Lebensweise, wirtschaftliche Bedeutung und systematische Kennzeichnung. *Zeitschr. angew. Entom.* VI, No. 2.

(2) BRITTAIN, W. H. (1915). *Hydroecia (Gortyna) micacea* as a garden pest. *Proc. Entom. Soc. Nova Scotia, Truro*, No. 1.

(3) BRITTAIN, W. H. (1920). Notes on the life history, habits and control of the pea moth. *Proc. Entom. Soc. Nova Scotia, Truro*, No. 5.

(4) BUCKLER, W. (1885–99). *The Larvae of British Butterflies and Moths*. 9 vols. Ray Society.

(5) CHITTENDEN, F. H. (1926). The common cabbage worm and its control. *U.S. Dept. Agric. Farmers' Bull.* No. 1461.

(6) COLE, A. C. and IMMS, A. D. (1917). Report on an infestation of antler moth. *Journ. Board Agric.* XXIV, No. 5.

(7) CURTIS, J. (1860). *Farm Insects*. Blackie and Son.

(8) FAHRINGER, J. (1922). Contributions to the knowledge of the habits of some Chalcids. *Zeitschr. wiss. Insektenbiol.* XVI, Nos. 11, 12; XVII, Nos. 1–4.

(9) FAHRINGER, J. (1922). Contributions to the knowledge of the habits of some parasitic Hymenoptera with special regard to their importance in the biological control of injurious insects. *Zeitsch. angew. Entom.* VIII, No. 2.

(10) FAURE, JEAN C. (1926). Contribution à l'étude d'un complex biologique— la Pieride du chou (*P. brassicae*) ét ses parasites Hymenoptères. *Lyon, Faculté des Sciences de l'Université.*

(11) FLUKE, C. L. (1920). The pea moth; how to control it. *Wisconsin Agric. Exp. Station, Madison, Bull.* No. 310.

(12) FRYER, J. C. F. and STENTON, R. (1923). Notes on the control of cutworms by poisoned bait. *Ann. App. Biol.* X, No. 2.

(13) GIBSON, A. (1915). The army worm. *Dom. of Canada, Dept. Agric. Entom. Div. Bull.* No. 9.

(14) GUNN, D. (1917). The small cabbage moth (*P. maculipennis* Curt.). *Union S. Africa Dept. of Agric. Pretoria, Bull.* No. 8.

(15) HARPER GRAY, R. A. (1915). Prevention of egg-laying on turnips by the diamond-back moth. *Journ. Board Agric.* XXII, No. 3.

(16) HEROLD, W. (1923). Zur Kenntnis von *Agrotis segetum* Schiff. III. Feinde und Krankheiten. *Zeitsch. angew. Entom.* IX.

(17) LEONARDI, G. (1922, 1925). *Insetti Dannosi e Loro Parassiti in Italia Fino al* 1911.

(18) MARSH, H. O. (1917). The life history of *Plutella maculipennis* Curt., the diamond-back moth. *Journ. Agric. Res.*, X, No. 1.

LEPIDOPTERA 93

(19) McDougall, Stewart (1915). Insect pests in 1914. *Trans. High. Agric. Soc. Scotland*, XXVII.

(20) McDougall, Stewart (1929). Insect pests. No. 5. Butterflies and moths. *Scott. Journ. Agric.* XII.

(21) Miles, H. W. (1926). Life history and control of the pea moth. *Bull. Chamb. Hortic.* III, Part 1.

(22) Miles, H. W. (1924). The diamond-back moth (*P. maculipennis* Curt.). *Ann. Rept. Kirt. Agric. Instit. Kirton, Lincs.*

(23) Ministry of Agriculture and Fisheries. Leaflets: No. 33 *Cutworms*, No. 22 *Diamond-back Moth* and No. 109 *The cabbage moth*. 10 Whitehall Place, London.

(24) Ormerod, E. (1890). *A Manual of Injurious Insects.* Simpkin, Marshall, Hamilton and Kent, London.

(25) Ripper, Walter (1928). Die Raupe der Kohlschabe (*Plutella maculipennis* Curt.). *Zeitsch. Insektenbiol.* XXIII.

(26) Seamans, H. L. (1926). *Pale Western Cutworm.* Dominion of Canada, Pamphlet 71, N.S.

(27) Smith, Kenneth M. (1926). Cabbage butterflies. *Gardeners' Chronicle*, Nov. 20.

(28) Somerville, W. (1916). A caterpillar on the ears of wheat. *Journ. Min. Agric.* XXIII.

(29) Theobald, F. V. (1925). *Cultivation, diseases and insect pests of the hop crop.* Min. Agric. Miscell. Publ. No. 42.

(30) Theobald, F. V. (1928). The large cabbage white butterfly and a simple method of control. *Journ. S.E. Agric. Coll. Wye, Kent*, No. 25.

(31) Theobald, F. V. (1903). The rosy rustic (*Hydroecia micacea*) attacking potatoes. *First Report on Economic Entomology.* British Museum.

(32) Theobald, F. V. (1928). The rosy rustic moth as a hop pest. *Journ. S.E. Agric. Coll. Wye, Kent*, No. 25.

(33) Theobald, F. V. (1926). The diamond-back moth. *Journ. Kent Farmers' Union*, XX, No. 3.

(34) Torka, V. (1929). Parasiten der Kohlschabe *Plutella maculipennis* Curt. *Anz. Schädlingsk.* V, No. 3. Berlin.

(35) Tullgren, Alb. (1926). Gräsflyet. En erinran i anledning av årets gräsmaskshärjningar. *Centralanstalten för Jordbruksförsok. Flygblad*, No. 115.

(36) Twinn, C. R. (1923). Studies in the life history, bionomics and control of the cabbage worm in Ontario. *54th Ann. Rept. Entom. Soc. Ontario.*

(37) Warburton, C. (1924). Ann. Rept. of the Zoologist for 1924. *Journ. Roy. Agric. Soc. England*, LXXXV.

(38) Zorin, P. V. (1924). The biology of *Phlyctaenia forficalis* L. *Plant Protection*, I, Nos. 1–2 (in Russian). Leningrad.

CHAPTER VIII

COLEOPTERA

Order 19. **Coleoptera.** Beetles

An extensive group, the members varying in size from very small to very large. The first pair of wings are horny, forming wing covers or elytra, these meet over the back in a mid-dorsal line, the *suture*, giving the insect a wingless appearance; hind wings membranous, folded under the elytra, not always present; mouth parts biting; metamorphosis complete.

Family *Cryptophagidae*

Very small beetles, seldom exceeding 3–4 millimetres in length; more or less elongate in form; first abdominal segment longer than the others; usually yellowish or brown in colour; antennae clubbed.

Atomaria linearis Steph. Pigmy Mangold Beetle

DESCRIPTION. *Adult.* Colour variable from deep red to black: elongate, parallel-sided. Head finely punctate; antennae brown, the three apical segments dilated. Thorax rather long and broad, finely punctured, base marginated. Elytra with short pale pubescence. Legs reddish brown. Length 1–2 mm.

LIFE HISTORY. The winter is passed in the adult stage; in autumn large numbers of the adult beetles may be found under clods of earth or debris in fields and meadows. These hibernating individuals emerge in May or June when they attack young mangolds and sugar-beet, just below soil level. The larvae do not appear to have been described; they probably feed upon the roots of the same host plants. It is thought that there are two generations a year.

HOST PLANTS. Chiefly mangolds, but it is also recorded from sugar-beet, of which it is a severe pest on the continent.

SYMPTOMS OF ATTACK AND INJURY TO HOST PLANT. The beetle attacks the young seedling plant just below soil level, producing an injury somewhat similar to that attributed to springtails (p. 21) except that the latter is above soil level. Injury is also caused to the roots, and a certain amount of feeding takes place upon the epidermis of the leaves. The most serious damage, however, is that below ground, as seedlings so attacked are usually killed, often breaking off in the wind.

DISTRIBUTION. The following remarks on the incidence of this beetle are quoted from Fryer (report on insect pests of crops for 1925–7). "In 1925 considerable damage was done locally in an area extending from the East Coast, south of the Wash to Bristol and including the southern counties. By 1926 the outbreak had returned to normal proportions, except in the S.E. province. A recrudescence of the pest to a limited extent occurred in 1927, considerable damage being caused to mangolds and sugar-beet over an area extending roughly from Nottingham and Derby to East Sussex."

CONTROL. Heavy rolling with a Cambridge roller has been found beneficial. Chemical treatment of the soil is not considered feasible owing to the difficulty of penetrating deeply enough. In the sugar-beet fields on the continent trapping is practised, holes are dug in the soil about 16 in. wide and deep, and these are filled with beet leaves and stems. Sowing crude naphthalene with the seed, at the rate of 1 part by weight to 20 parts of beet seed, is also practised as a deterrent. Attacks on sugar-beet are always worst when this crop follows beet or mangolds, they are less serious after swedes or cabbage.

Family *Scarabeidae*. Chafers; Dung Beetles, etc.

A large group of beetles comprising about 14,000 species, of which 90 are found in the British Isles. They are mostly large insects, convex in shape with the terminal segments of the antennae in the form of flattened plates which are capable of being brought close together. Their habits divide the group roughly into two divisions, the dung feeders and the chafer beetles, the latter only being of direct agricultural importance.

Melolontha melolontha F. (*vulgaris* F.). The Cockchafer

DESCRIPTION. *Adult.* A large, rather convex insect, the general body colour reddish brown. Head black with white transversely directed bristles; clypeus prominent, red with a raised margin and black rim; antennae reddish, 10-jointed, the third longer, a lobed fan on the terminal segment. Thorax broad, black with white bristles; scutellum rounded, black with a fringe of yellowish hair; Elytra reddish brown with five raised lines and short white pubescence; abdomen produced into a spade-like point. Legs reddish with black markings.

In the male the antennal club is long and consists of seven plates; in the female, it is short and consists of six plates only. Length 22–26 mm.

Egg. Roundish, white or pale yellow.

Larva. Large and fleshy, white or creamy white in colour, curved in a characteristic manner; head large, brown, shining; mandibles large with a granulated area where the light and dark parts meet; antennae 4-jointed, a little longer than the mandibles; legs pale brown, first pair shorter than the other two pairs which are of equal length, segments transversely rugose, the last segment large and frequently swollen, appearing black from the contents of the intestine which are visible through the skin; dorsally there are a number of short bristles with some longer ones intermingled. Length 48 mm. (Fig. 20.)

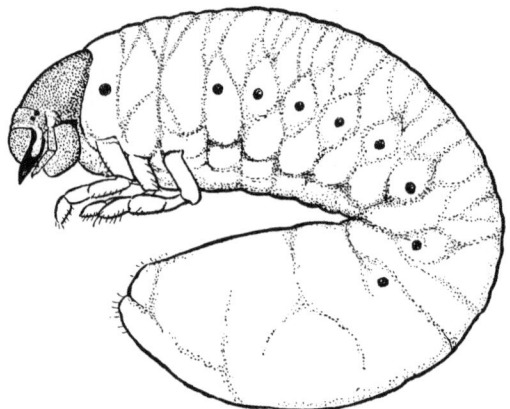

Fig. 20. Larva of *Melolontha melolontha* F. The Cockchafer. × 3.

The shape is very characteristic, the larva always lying in the soil with the body bent in the manner shown in Fig. 20. This larva rather closely resembles that of *Geotrupes*, the dor beetle, but the latter can be readily distinguished from the chafer larva by the third pair of legs which are reduced in size to act as a stridulating organ against the second pair.

Pupa. Rather large, the abdominal segments slightly curved.

LIFE HISTORY. The adult beetles emerge from the soil at the end of April or early in May; mating takes place a few days after emergence but the eggs are not ready for deposition until three weeks later. They are deposited in the ground at a depth of about 6–8 inches in batches of 12–30. The larva hatches in 3–4 weeks and lives in the soil for a period of three years in the British Isles and for longer or shorter periods in other countries according to the climate. In France it is found that in cold districts the life cycle takes four years instead of three, and even in regions where the normal cycle is three years, exceptional climatic conditions may result in the production of a four-year generation. During the first year of the larval life little damage is done to crops, but later the larvae feed near the surface on the roots of cereals, grass and young trees, descending to a greater depth during the winter. At the end of the third year, pupation takes place at a depth of about 2 ft. The adult emerges from the pupa in October but remains in the pupal cell until the following spring, the fourth year from commencement of development.

HOST PLANTS. The adults feed on the leaves of various trees, particularly oaks, while the larvae feed on almost any roots, but chiefly those of grass and seedling trees. They have also been recorded from rhododendrons, strawberries, especially when grown near recently cleared woodlands, potatoes, cereals and hops.

DISTRIBUTION. Very common all over the British Isles and most of Europe.

CONTROL. This is a difficult pest to control; naphthalene is recommended against attacks by the larvae on grassland, used at the rate of 2 cwt. per acre; it can be applied in either autumn or spring and is most effective if followed by rain. Paradichlor-benzene is also worth trying; apply in holes about 1 foot apart, 1 gramme to each hole. In America some success has been achieved by the use of lead arsenate applied at the rate of 5 lb. lead arsenate to a bushel of moist screened sand or soil and broadcast over each 1000 sq. ft. of turf. Pigs and poultry turned on to infested land after deep ploughing will reduce the numbers. Collection of the adults, if possible before oviposition, is also much to be recommended. Theobald found that steam-rolling infested pasture in October was very successful.

NATURAL ENEMIES. Birds such as rooks, magpies, missel thrushes, green plover, and black-headed gulls prey upon the larvae. The Tachinid fly *Dexia rustica* F. is recorded from *M. melolontha* in several European countries and the fungi *Botrytis bassiana* and *B. tenella* in Italy, but little seems to be known of the insect enemies. Scheidter(53) considers that the great increase in numbers of these beetles is probably due in part to the fact that all stages except the adult are well protected in the soil against enemies.

Amphimallus (Rhizotrogus) solstitialis Linn. Summer Chafer

DESCRIPTION. *Adult.* General colour brownish-red. Head dark; clypeus red with a raised black margin; antennae reddish 9-jointed, club with three lamellae, third and fourth segments about equal in length. Thorax darker with long reddish bristles; scutellum thickly clothed with pale hairs. Elytra reddish brown shining, with four rather indeterminate raised lines and some black markings, very few bristles, extremity of abdomen not produced. Legs rather long, reddish brown, outer side of fore tibiae rimmed with black.

In the male the antennal club is larger; female with three projections on the external face of the anterior tibiae, thorax less villose. Length 14–16 mm.

Larva. Very similar to the larva of the foregoing species, but considerably smaller when both are mature; it can be distinguished from the larva of *melolontha* by the under surface of the mandibles, which is minutely granulated. Length 24–30 mm.

LIFE HISTORY. The habits of this insect are very similar to those of *melolontha*, the larvae, however, seldom attack young trees but are rather pests of agricultural crops. The life history is probably completed in one year, though no study appears to have been made upon the bionomics of this species and so this cannot be definitely stated. The winter is spent in the larval condition, pupation taking place in the spring. The beetles are nocturnal, flying from about seven o'clock in the evening onwards. The same control methods are applicable to this species as are given for *melolontha*.

NATURAL ENEMIES. The following are recorded from Europe, the Scoliid *Tiphia femorata*, and the Tachinids *Billaea pectinata* Mg., *Dexia rustica* F. and *Syntomocera petiolata* Bonsd.

Phyllopertha horticola Linn. Garden Chafer, 'Bracken Clock'

DESCRIPTION. *Adult.* Rather flattened with a greenish metallic shine. Head anteriorly rugose, posteriorly with large punctures; antennae reddish, club black. Thorax narrower than elytra, coarsely punctured; scutellum black. Elytra reddish brown, rather shiny with dark suture and some indeterminate lines. Legs black, male with one claw dilated. Length 7–10 mm.

Larva. The smallest of the four species of chafers here dealt with; abdomen rather more cylindrical than the foregoing species but still somewhat clubbed, third joint of antennae of the same length as first joint; tibiae half the length of the femora; claws of legs increasing in size; spiracles round and flat. This larva can be distinguished by the character of the mandibles which have on them a pale oval area with file-like ridges across it. (Fig. 21.) Length 18 mm.

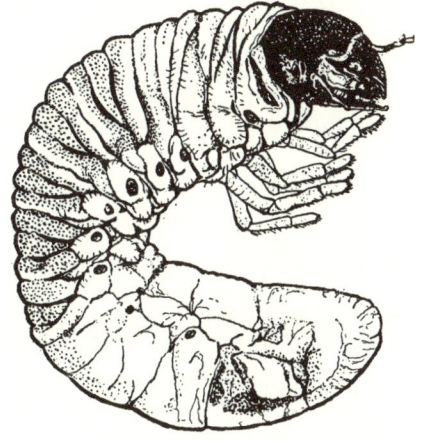

Fig. 21. Larva of *Phyllopertha horticola* Linn.
The Garden Chafer. × 5.

LIFE HISTORY. The adult beetles appear about the second week of June and are most numerous during the latter part of that month, they may be found feeding upon *Rumex*, clover and other broadleaved plants; the most favoured food, however, is the young shoots of bracken. The eggs are deposited in the soil, usually of grassland, and the resulting larvae live at a depth of 1–6 in.; the larval life occupies about eight months until the following May, when pupation takes place. The pupa is found about 4 in. below the surface, enveloped in the cast-off larval skin. The larvae are chiefly injurious to farm crops rather than garden plants and are often a severe pest of pasture land, the roots of the grass being destroyed. The tendency of such attacks on pasture, if occurring for several years in succession, appears to be to remove the fine leaved types of grass and so reduce the quality of the sward. Attacked areas of grassland can be recognised by the soft and spongy texture of the ground.

CONTROL. Damage appears to be less where attempts at grass-land improvement have been made. An autumn dressing of basic slag followed by a dressing of kainit the ensuing May is to be recommended and harrowing and reseeding of bared patches. Control by insecticides is impracticable. A dust of the modern insecticides might be devised for controlling the beetles on a large scale during the comparatively short flight period. (Harper-Gray, Peet and Rogerson (1947), *Bull. Entom. Res.* XXXVII, 455–68.)

Cetonia aurata Linn. Green Rose Chafer

DESCRIPTION. *Adult.* A characteristic and unmistakable insect; bright golden green, shining. Head rugose or coarsely punctured;

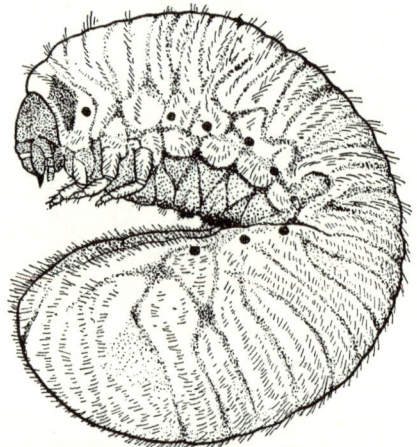

Fig. 22. Larva of *Cetonia aurata* Linn.
The Green Rose Chafer. × 3.

antennae dark, club with three plates. Thorax narrow anteriorly with few large punctures; scutellum triangular. Elytra flattened in centre with a raised line on each side of the suture, some long bristles, punctures of a horseshoe-shape; lower part of elytra with transverse, white, sinuate markings. Length 14–20 mm.

Larva. Fat and whitish, head ochreous, legs reddish ochreous, body clothed with transverse rows of reddish hairs. In size and general appearance the larva much resembles that of *Melolontha melolontha*, the cockchafer, but differs from it in the length of the legs which are shorter with pointed claws, by the possession of a brown horny spot on each side of the segment next the head and by the regular transverse rows of short ferruginous hairs on the body. Length 24–36 mm. (Fig. 22.)

LIFE HISTORY. The eggs are usually deposited in the soil where the larvae feed for two or three years before pupating. When mature, the larvae form a cocoon, about the size of a walnut, covered externally with particles of soil. The adults appear in May and June and feed upon flowers, particularly of rosaceous plants. Ormerod records them as destroying turnip blossom. The larva is often a severe pest of pasture land, and the Ministry of Agriculture record the destruction by it of four acres of pasture in south Breconshire in 1925, a similar attack in the same field having been reported in 1921. The control methods recommended for the larva of the cockchafer apply also to this species.

NATURAL ENEMIES. The Tachinid *Billaea pectinata* Mg. is recorded in Europe from this beetle.

CHIEF CHARACTERISTICS OF THE LARVAE OF THE FOUR SPECIES OF CHAFER BEETLES

Melolontha melolontha. Cockchafer

Mandibles with a granulated area; third joint of antennae two thirds the length of first joint; tibiae three quarters length of femora; claws of legs diminish in size; anterior spiracles oval, posterior spiracles round; legs rather long. Length 24–36 mm. (Fig. 20.)

Cetonia aurata. Rose Chafer

Similar in size and general appearance to above, differentiated by the short legs, pointed claws, the possession of a brown horny spot on each side of the segment next the head and the transverse rows of short hairs. Length 24–36 mm. (Fig. 22.)

Amphimallus solstitialis. Summer Chafer

Mature larva smaller than the foregoing, the whole surface of the mandibles very minutely granulated. Length 18–24 mm.

Phyllopertha horticola. Garden Chafer

The smallest of the four species; on the mandibles is a pale oval area with file-like ridges across it; third joint of antennae the same length as the first joint; tibiae half the length of the femora, claws increasing in length; spiracles round. Length 18 mm. (Fig. 21.)

Family *Elateridae*. 'Click Beetles'

Mostly elongate beetles; antennae pectinate, serrate or filiform; prothorax with the hind angles generally produced; many of the members of this group possess the power of leaping when lying on their backs.

The larvae of certain of the *Elateridae* are known as 'wire-worms' and constitute some of the worst agricultural pests. Four species are dealt with here, three of them belonging to the genus *Agriotes* and the fourth to the genus *Athous*. The description and life history of each species are given separately, but the host plants, damage and control are dealt with collectively for all four species. The group has been extensively studied by Rymer Roberts(52) from whose work much of the information given here has been obtained.

Agriotes lineatus Linn.

DESCRIPTION. *Adult.* Oblong, brown. Head triangular, thickly and rather coarsely punctured; palps and antennae light brown.

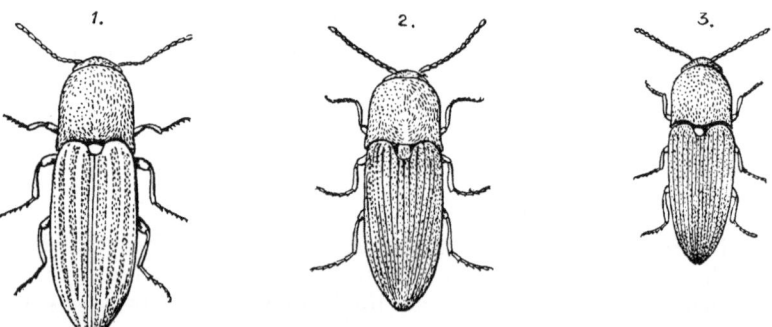

Fig. 23. *Agriotes* spp. Click Beetles. 1, *A. lineatus*; 2, *A. obscurus*; 3, *A. sputator*. All much enlarged. (Reproduced by permission of the Ministry of Agriculture.)

Thorax fuscous, darker with a scanty brown pubescence, thickly punctured, posterior angles produced. Elytra regularly striated with coarse punctures, the interstices with brown pubescence, giving a striped appearance. Legs brown, femora slightly darker. Length 8 mm. (Fig. 23, 1.)

Egg. Opaque, milky white, nearly spherical, but slightly longer than broad. Length about 0·5 mm.

Larva. The larvae of the three species of *Agriotes* are similar in general appearance but differ in certain microscopical characters which will be described. They are elongate, yellow or yellow-brown in colour, with a very resistant and shiny integument. The head is dark with large mandibles; legs short and simple. The principal distinguishing characters of the *Agriotes* larvae are, firstly, the presence of a tooth situated dorsally on the inner edge of the mandibles near the apex and, secondly, the presence of two eye-like pits situated near the base of the ninth abdominal segment, and referred to, probably erroneously, as 'sensory pits'. The larva of *A. lineatus* may be differentiated from the larvae of the other species of this genus by the shape of the spiracles which are longer and more parallel-sided and which also bear a larger number of teeth bordering the orifice.

Fig. 24. Pupa of *Agriotes* sp. (Much enlarged.)

Pupa. The pupa is free, pale, almost white in colour and soft. The general appearance of *Agriotes* pupae is shown in Fig. 24.

LIFE HISTORY. The adult beetles hatch in late summer and may be found occasionally during the winter in tufts of grass and hedge bottoms; probably also a number, although hatched, remain in the soil till the following spring. About the middle of May the beetles emerge, oviposition taking place in June and July. The female burrows in the soil and deposits her eggs at a depth of $\frac{1}{4}$–2 in. below the surface; if laid too near the surface they are prone to dry up, the wireworm in all its stages being particularly prone to desiccation. The eggs are deposited either singly or in clusters, mostly at the roots of grasses, in pasture, temporary ley or badly cultivated land containing couch grass and other grass weeds. The larvae hatch in about four weeks, they are pale, almost white and semi-transparent, the colour gradually deepening to the characteristic yellow as growth proceeds. During the first year the food consists mainly of decomposed vegetable matter; growth during this period is slow, the first moult taking place in the following June. After this, growth is more rapid, the second moult taking place in July or August, and then subsequently twice a year, the first in April or May, the second between July and September; the total life cycle of the *Agriotes* larvae being probably five years. Pupation takes place in the soil at a depth of about 6 in., in an earthen cell prepared by the larva, the pupa being erect, head uppermost. The pupal period lasts about 3–4 weeks.

Agriotes obscurus Linn.

DESCRIPTION. *Adult.* Darker, more rounded than *A. lineatus*; antennae and palps dark brown. Thorax thickly and strongly punctured. Elytra convex without the lined appearance of *lineatus*, somewhat dilated at the middle and narrowing towards the apex. Legs dark brown or yellowish, femora darker. Length 8–9 mm. (Fig. 23, 2.)

Larva. Head and ninth abdominal segment somewhat flattened, remainder of body nearly cylindrical; yellow, usually rather pale; junction of the body segments rather darker than the rest of the body and marked with faint longitudinal striae; anal segment terminated by a slight constriction which swells out into a blunt point. Length 22–26 mm.

Pupa. General colour white, speckled on the apices of the abdominal segments with small patches of rusty brown; thorax nearly as broad as long and swollen. Two spines project from the anterior angle of the thorax; the third and sixth abdominal segments are slightly tooth-shaped at their lateral posterior margins. The ninth abdominal segment is provided ventrally with a flap, broad at the base, suddenly narrowing and terminating in two sharp points. (Ford (14).)

LIFE HISTORY. When mature the wireworm makes a cell about 3–4 in. below the surface. It then pupates, and by the end of September most of the pupal cells contain young adults which remain there until the following March. They then emerge but do not commence egg-laying till April. During this month, May and early June, the eggs are laid in batches in the surface soil, either in damp places or during showery weather. The eggs are nearly spherical and about 1/50th in. in diameter. One beetle lays 40–100 or more eggs. They take 5–6 weeks to incubate and young wireworms hatch in June and July (Miles (1942), *Ann. Appl. Biol.* XXIX, 176–80). Ford considers the length of the larval life of this species to be four years but Rymer Roberts puts it at five years.

Agriotes sputator Linn.

DESCRIPTION. *Adult.* The smallest of the three species of *Agriotes*; general colour black or dark brown with whitish pubescence. Head rather rounded; antennae brown or yellowish, second joint larger than the third and fourth. Thorax convex, shining, fairly thickly and finely punctured, lighter in colour than in *A. lineatus*, posterior angles rather sharp. Elytra with rather deep striae, interstices rugose, grey pubescence. Legs brown. Length 5–6 mm. (Fig. 23, 3.)

Larva. Very similar to that of *A. obscurus*, but usually shorter and darker in colour. Length 16–17 mm.

LIFE HISTORY. The adults of this species together with *A. obscurus* may often be found in great numbers under heaps of cut grass. The length of the larval period is probably four years; in most respects the habits of this species are similar to those of the other species of *Agriotes*.

Athous haemorrhoidalis Fab.

DESCRIPTION. *Adult*. Distinguished from *Agriotes* by its larger size and reddish colouring; elongate. Head dark, often black; antennae dark, rather long, second joint shorter than the third; head thickly and coarsely punctured. Thorax dark, longer than broad, sparsely punctured, posterior angles rather short. Elytra brown to reddish with punctured striae, interstices rather finely punctured, suture black, under side of abdomen red. Legs dark brown. Length 9–12 mm.

Larva. Biconvex, flattened, deep yellow, strongly chitinised on the dorsal surface; head, prothorax and ninth abdominal segment, brownish yellow. This species is easily distinguished from the *Agriotes* larvae by its broader and more flattened appearance, and by the presence on the dorsal surface of the ninth abdominal segment of a kind of impressed shield, bearing at its posterior end two pairs of prongs or cerci which give it the appearance of possessing a bifurcated tail. This shield is present in other Elaterid larvae, but the orange-yellow colour of *A. haemorrhoidalis* together with a single median furrow, from which 3–4 transverse furrows branch, on the ninth abdominal segment, will serve to distinguish it from most other Elaterid larvae. This larva also lacks the mandibular tooth and the eye-like pits on the ninth abdominal segment which are characters of *Agriotes* larvae.

LIFE HISTORY. The adults remain in the pupal cells during the winter, they emerge about the middle of May and occur until July. The beetles are often found on leaves and flowers, especially the flowers of *Caucalis anthriscus*, hedge parsley, and on growing corn, clover and hedgerow plants. So far as is known the life history is similar to that of *Agriotes*.

HOST PLANTS OF WIREWORMS. In his *Farm Insects*, Curtis refers to the wireworm as an "example of a larva which may be termed omnivorous as regards the production of field and garden" and it is true that wireworms occur in many kinds of soil and attack almost every kind of crop. It will be best therefore to indicate the class of crop most attacked by each species of wireworm without attempting to give a full or comprehensive list of food plants. The larvae of *A. lineatus* attack field crops, vegetables, ornamental plants and forest seedlings; *A. obscurus* attacks cereals, root crops, fodder and

ornamental plants; *A. sputator* is a pest of carrots and other vegetables, and to a less degree of potatoes, and may attack sugar-beet.
Athous haemorrhoidalis is more commonly found as a pest of
potatoes, but also attacks cereals and grasses. All three species of
Agriotes larvae will feed upon grasses, which were probably the
original food plants, and also many weeds such as couch grass (*Agropyron repens*), charlock (*Sinapis arvensis*), and dock (*Rumex* sp.),
the last named being a favourite host of *Athous haemorrhoidalis* as
well. Mustard and rape, though not immune, are only attacked in
the absence of some more palatable crop. The larvae of *A. haemorrhoidalis* seem to prefer pastures and meadows to arable land; they
are often fairly numerous along hedgerows, feeding amongst the
roots of grasses and hedgerow plants. *Agriotes* larvae are able to
live for as long as 20 months in soil in the absence of any crop,
feeding on the humus; at times also they may develop the carnivorous habit.

INJURIES TO HOST PLANT AND SYMPTOMS OF ATTACK. Injury by
wireworms varies somewhat, according to the nature of the host
plant; in most cases the damage is caused to the roots and underground portions of the plant. Plants like onions or celery are
attacked at the fleshy collar, causing them either to die or go prematurely to seed, while root crops and potato tubers are riddled
with holes. In the case of beans and tomatoes the wireworms enter
the stems at soil level and bore upwards above ground, peas
are usually riddled in the soil and fail to germinate. Cereals
are attacked in the young stage, the larvae passing from plant to
plant; young wheat plants are usually attacked at the interval
between the seed and the bulb, the stem at this point is often bitten
through, causing the whole plant to wither, and not merely the
centre shoot as is the case with the larva of the wheat bulb fly
(p. 212). The plant becomes loose in the soil, shrivels and is easily
pulled out or blown out by the wind. A characteristic of wireworm attack on cereals is the death of a number of consecutive
plants in a row.

DISTRIBUTION. The distribution of the various wireworms is
localised; what is the common species in one district may be scarce
in another. In Cheshire, Lancashire and Shropshire the species
usually met with is *Agriotes obscurus*. This is indeed, perhaps, the
commonest species, being generally distributed in the United
Kingdom. *A. lineatus* and *A. sputator* also occur but become
scarcer further north. *Athous haemorrhoidalis* is perhaps the least

COLEOPTERA 107

common of all, although the adult, because of its more active movements, is often encountered.

CONTROL. Attempts to control wireworms are best divided into two classes; the first or cultural method aims at reducing the numbers of wireworms by suitable cultivation, while the second class consists of more direct methods such as soil insecticides, poison baits, trap crops, etc. As regards cultural methods, wireworm attack is most severe upon crops following after grass, leys or foul land. Leaflet No. 10 of the Ministry of Agriculture recommends that such land should be broken up in the latter half of the summer, and in the case of meadow land, directly after the hay crop has been carted, so as to make a bastard fallow; this method also checks the leatherjacket (p. 157). Both pasture and grain stubble should be ploughed early with a good strong furrow so that all is buried with the sod. Thorough cultivation of the ground is an important point; it has been found that oats, when sown under proper conditions with a good tilth, grew away from wireworm attack, while oats sown in badly ploughed land with some furrows still unbroken failed to withstand attack. Where land likely to be already infested, such as pasture or leys, is to be broken up, it is possible to avoid loss by careful attention to the succeeding crop. Lees (35) has divided agricultural crops into classes according to their degree of susceptibility to wireworm attack. He places in the most susceptible class plants attacked at the fleshy collar and which are killed, dwarfed or caused to go to seed prematurely, such as onions, leeks, celery and lettuce. Less susceptible are leguminous crops in which the plant is dwarfed but not killed, while slightly susceptible are Brassicas and potatoes, to which may be added rape and mustard as the least susceptible of all. If mustard is to be grown on infested land, it may be treated in one of three ways; it can be sown in April or May as a seed crop and allowed to occupy the land the whole summer; it can be sown in the summer and ploughed in when about 18 in. or 2 ft. high, or it can be sown in the summer and eaten off by sheep. Mustard oil is apparently highly toxic to wireworms. As regards the second class of remedial measures, where more direct action is contemplated, there is first the application of chemicals to the soil. This method is not always successful owing to the resistant nature of the wireworm, but heavy dressings of crude naphthalene, kainit alone or mixed with quicklime have sometimes proved effective. Secondly, dressing the seed with chemicals previous to sowing has been sometimes successful;

other methods consist of trap crops, heavy rolling, baits such as heaps of clover or lucerne poisoned with arsenic and the application of soil insecticides. Some recent work in Germany (Blunck and Merkenschlager(3)) has shown in the course of soil analyses from fields infested with *Agriotes* that the feature common to all cases was the assembly of larvae in places where the soil had the lowest percentage of alkali or other bases, and even a marked degree of acidity did not repel them. Recent work in this country and in America with a combination of trap crop and soil insecticide has given good results and is here shortly described. Briefly, the method consists of attracting the wireworms to one portion of the field by means of a suitable trap crop and then destroying the assembled wireworms with heavy doses of calcium cyanide. This method has recently been tested in England by Miles and Petherbridge(42) who baited portions of infested fields in autumn with trap rows of wheat and oats or wheat and bran. They found that the wireworms assembled in about 14 days, or longer in cold weather. The bait rows must be examined at intervals to determine whether the wireworms are assembling. The calcium cyanide should be covered with soil immediately after application and the land rolled or pressed to close the larger air spaces and prevent too rapid escape of hydrocyanic acid gas. On agricultural land with wheat as bait drilled in rows about 3 ft. apart, $3\frac{1}{2}$ stone is sufficient for an acre and about 90 lb. of calcium cyanide would be required, costing about £4. 10s. This cost is high but an advantage of the baiting system is that by examining the bait it is possible to locate the main areas of wireworm infestation and only the infested portions of the field need be treated with the calcium cyanide. Six or seven days after treatment the land can be lightly cultivated and plants set or seed sown within the next few days.

In 1941 a symposium on wireworm investigations was held in London and published in the *Annals of Applied Biology* for 1942. From this the reader can obtain a fair idea of the situation up to that time. Some of the salient points from the symposium are quoted here. From the agricultural aspect, the importance of wireworms arises from three well-known facts. First, wireworms, especially *Agriotes*, are essentially inhabitants of a mixed plant population in which members of the Gramineae predominate. Secondly, wireworms spend a long time—something like five years

—in the larval stages. Thirdly, wireworms are able to feed on a wide range of important crops, notably cereal crops. From this it follows that the wireworm problem is likely to be serious in any system of rotation in which grassland alternates with arable. The importance of wireworms as pests, turns therefore, on the trend of agricultural practice; when this is to lay down arable to grassland their importance diminishes, while it increases when the reverse takes place.

As regards the control of wireworms it was suggested that by the use of measures already known, e.g. by the choice of suitable crops and methods of cultivation, the loss from wireworms could be greatly reduced provided the measures were selected in relation to the dimensions of the wireworm populations present. A scheme was therefore drawn up, having three objects:

(1) To find a method of estimating the wireworm population of a grass field with sufficient accuracy and sufficient speed.

(2) To obtain some guide as to the damage to be expected from wireworm populations of different sizes.

(3) To discover any obvious correlations between the size of the wireworm population in any field and other factors, such as soil type or cultural treatment (Fryer (1942), *Ann. Appl. Biol.* XXIX, 144-9).

The discovery of the new insecticides, D.D.T. and Gammexane may lead to more direct methods of attack and preliminary experiments with these two substances have given promising results. As regards Gammexane, some experiments seem to suggest that the application quickly renders wireworms incapable of attacking the crop but any killing action appears to be considerably delayed. (Dunn, Henderson and Stapley (1946), *Nature, Lond.*, CLVIII, 587.)

NATURAL ENEMIES. Birds are of considerable benefit in clearing infested land, particularly the black-headed gull, poultry and pheasants. Not many Hymenopterous parasites are known; Rymer Roberts records a Proctotrypid bred from *Athous haemorrhoidalis* and Régnier(48) a Proctotrypid also, *Phaenoserphus pallipes* Latr., from *A. obscurus*. Hodson in England and Blunck in Germany record the Proctotrypid, *Paracodrus apterogynus* Hal. from *A. obscurus* and *A. sputator*. The fungus *Empusa carpenteri* is recorded from France.

CHAPTER IX

COLEOPTERA (*contd.*)

Family *Chrysomelidae*. Leaf Beetles

Short compressed beetles, generally very convex, upper surface usually bare and shining with metallic colours; antennae of moderate length, points of insertion distinct from the eyes.

Lema melanopa Linn. (*Crioceridae*)

DESCRIPTION. *Adult.* Head and eyes black; antennae black with the first two segments rather globular. Thorax light brown, sometimes darker, shining. Elytra shining, blue with parallel perpendicular lines of punctures. Legs yellow, with tarsi and tips of tibiae black. The males appear to be slightly smaller and narrower than the females. Length 4–4½ mm. (Fig. 25.)

The following descriptions of the immature stages of *L. melanopa* are given according to Hodson (22).

Egg. Cylindrical, rounded at ends, shining yellow and covered with a glutinous secretion which hardens on contact with the air. Colour darkening almost to black, prior to hatching. Length 1 mm.

Larva. Eruciform; head, spiracles and legs somewhat heavily chitinised, dark brown to black; remainder soft, more or less

Fig. 25. *Lema melanopa* Linn. Adult beetle. × 10. (After Hodson.)

wrinkled, dirty to bright yellow, bearing numbers of stiff curved hairs which in the dorsal regions point forwards and on the sides upwards; there are nine pairs of spiracles; anus situated dorsally and directed forwards, in life the colour is usually obscured by a covering of excrement. Length 5–6 mm.

Pupa. Chrysomelid type, enveloped in a thin transparent membrane, bright yellow, darkening rapidly until normal adult colouring is attained. Length 4–4·5 mm.

LIFE HISTORY. Hibernation is probably only in the adult stage, in stacks, crevices in bark, flood refuse, etc. The beetles reappear about April, mating taking place towards the end of May. According to Hodson (22), who has studied this species, oviposition occurs about a week after mating, the eggs being deposited singly on or near the midrib, on the upper and inner side of the leaves of cereals; oviposition continues from the end of May till the end of July. The larva hatches after about a fortnight and feeds upon the upper epidermis of the leaf, which it skeletonises in long strips. There are four larval instars; when mature the larva pupates in the soil where a pupal chamber is constructed at a depth of 2–3 in. The total life cycle occupies about 46 days and there is probably only one generation per annum.

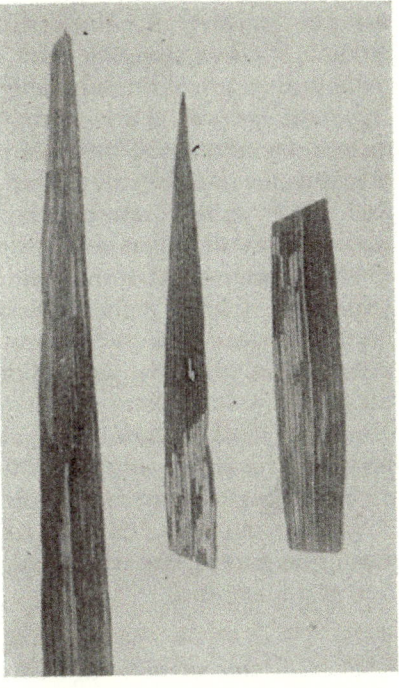

Fig. 26. Oat leaves showing typical injury by larvae of *Lema melanopa* Linn. About natural size. (After Hodson.)

CULTIVATED HOST PLANTS. Barley, oats and wheat.

WILD HOST PLANTS. Cocksfoot (*Dactylis glomerata*), timothy grass (*Phleum pratense*), couch grass (*Agropyron repens*), canary grass (*Phalaris canariensis*).

SYMPTOMS OF ATTACK AND INJURY TO HOST PLANTS. Both adults and larvae feed upon cereals, the chief damage being done by the adults in spring when food is scarce and feeding is confined mostly to the foliage of young corn, longitudinal strips are eaten from the leaves somewhat after the manner of slug damage. The larvae feed generally on leaves of cereals and still further increase the damage already done by the adults. (Fig. 26.)

DISTRIBUTION. First recorded in Great Britain as causing damage in 1917 from Shropshire and Kent; since that time it has been recorded from North and mid-Wales, Devon and Cornwall, Hertfordshire and Northumberland.

CONTROL. Many measures for control are recommended on the continent where *L. melanopa* is a serious pest; most of these measures, however, are unsuited to agricultural practice in Great Britain. Hodson suggests that investigation of the following methods may prove the suitability of some of them.

Cultural methods. Early sowing of spring corn is recommended, and also the premature cutting of areas attacked by the larvae; this latter practice is aided by the fact that attacks by *L. melanopa* are often patchy in their nature, being at first restricted to strips near headlands and the edges of the field. The cut plants should be dried rapidly or removed from the field.

Spraying. This is only practical in the case of small areas of infestation when some such poison as lead arsenate might be used, or that in use in Russia, i.e. sodium arsenite and lime (1 : 4) at the rate of 1 lb. to 96 gallons of water.

Dusting. This method being cheaper and more easily applied than a spray is worthy of a trial; Megalov[38] recommends a dust of one part of calcium arsenate to five parts of finely sifted lime.

NATURAL ENEMIES. Hodson records two Hymenoptera parasitising the larvae, one a Chalcid, *Tetrastichus* sp., the second at present unidentified.

Crioceris asparagi Linn. Asparagus Beetle

DESCRIPTION. *Adult.* Blue-black or greenish, head black, antennae dark brown, thorax rusty red with some black markings; elytra with outer margins yellow, inner margins black; there is a black line down the suture from which arise two black transverse bars giving the appearance of a cross. Length 6–7 mm.

Egg. Rather large in proportion to the adult, elongate-oval, dark brown in colour, about three times as long as wide. Length 2 mm.

Larva. Head shining black, body soft, fleshy, curved and much wrinkled; olive or slate grey in colour; legs black, abdominal feet present on each segment, there is a large anal proleg with which the larva clasps the plant. Length 6 mm.

Pupa. Yellowish in colour, enclosed in a parchment-like cocoon often covered with soil particles. Length 5 mm.

LIFE HISTORY. Egg-laying takes place from June onwards, at first upon the heads and shoots and later upon the feathery foliage of the asparagus plants. The eggs are deposited in rows of 3–7 or more, the larvae emerge after 5–7 days and feed for about a fortnight, there being probably three moults. Pupation takes place in the cocoon just below soil level, the pupal period lasting 14–20 days. There are two or three broods in the year, both adults and larvae occurring as late as October. Hibernation takes place in the adult stage under stones, in crevices in the soil, etc.

CONTROL. Lead arsenate sprays or dusts are usually recommended for the control of *C. asparagi*. The following sprays may be used: 1 lb. Urania green (Paris green), 10 lb. slaked lime and 100 gallons of water; or 4 lb. lead arsenate to 50 gallons of water. If a dust is used apply the lead arsenate mixed with hydrated lime or as an alternative a 2 per cent. nicotine dust is sometimes effective.

NATURAL ENEMIES. The following parasites and predators have been recorded as attacking *C. asparagi*: In France, the Tachinid fly, *Meigenia floralis* Meig., and the Sphegid *Cerceris quinquefasciata* Rossi; in America the Coccinellids *Megilla maculata* De G. and *Hippodamia convergens* Guér. are predacious upon the larvae, while the egg is parasitised by the Chalcid *Tetrastichus asparagi* Cwfd. (Johnston, F. A., *Journ. Agric. Res.* IV, 1915.)

Phyllodecta vitellinae Linn. The Willow Beetle

DESCRIPTION. *Adult.* Oval, bronze or greenish, shining, sometimes coppery. On the elytra are rows of *regular* striae. This insect is distinguished from the closely similar species, *P. vulgatissima*, by its bronze colour and the regularity of the striations. Length $3\frac{1}{2}$–$4\frac{1}{2}$ mm.

Larva. Dirty grey to yellowish grey with darker spots.

LIFE HISTORY. The winter is passed in the adult stage, the beetles congregating under the bark of willows and poplars and sometimes in rotten wood. In spring they migrate to the willow, where the female oviposits, the eggs being laid on the under sides of the leaves in batches of 6–12. The larvae feed gregariously in straight rows on the under surface of the leaves, often reducing them to skeletons. When full grown they descend into the soil and pupate; there are probably two broods in the year. Both larval and adult stages cause considerable damage to the osier beds; in addition to the injury to

SAE   8 the leaves caused by the feeding of the larvae, the beetles stunt the length of the rods by eating the tender growing points of the young rods. *P. vitellinae* attacks many species of willows, particularly *Salix purpurea* and *Salix alba vitellina*; according to Hutchinson and Kearns (23) *Salix triandra* is not attacked. A lead arsenate dust is recommended as a control for this insect.

Phyllodecta vulgatissima Linn. (*betulae* Linn.)
Allied Willow Beetle

DESCRIPTION. *Adult.* Oval, shining, metallic, generally bluish or bluish green. Head narrow, shining, punctate; antennae with first joint dilated, flattened, last joint oval. rather pointed. Thorax

FIG. 27. Larvae (× 5) and Adults (× 6) of *Phyllodecta vulgatissima* Linn. The Allied Willow Beetle. (After J. Keys.)

much broader than head, punctate. Elytra broader than thorax, punctate, *irregularly* striate, shining. Legs dark or black, metallic, with some pale pubescence on tarsi and base of tibiae. Male with posterior tibiae curved and first joint of tarsi strongly dilated. Length 3½–5 mm. (Fig. 27.)

Larva. Similar to that of *P. vitellinae*, the general appearance of the larvae of both these species is shown in Fig. 27.

The life history and habits of this species are the same as those of the foregoing *vitellinae*. It appears to be confined mostly to varieties of *Salix viminalis*.

Galerucella lineola Fab. Willow Beetle

DESCRIPTION. *Adult.* Oblong, dull brown, closely pubescent. Head brown, vertex black, eyes black, prominent; antennae black. Thorax densely punctured, pubescent except at margins, a large black spot on dorsum. Elytra long, rather narrow, parallel-sided, densely punctured, sometimes with a black mark at the apex of the suture and on each humerus. Legs reddish brown with base of tibiae and tarsi blackish. Length 4–5 mm.

Larva. The larvae are of a yellow-black colour with black spots and patches; under surface yellow with dark patches and dots; laterally are a number of tubercles. Length 12 mm.

LIFE HISTORY. Hibernation is in the adult stage, mostly under fallen leaves. In spring the adults emerge and attack the young shoots of willows. The orange-yellow eggs are deposited in clusters on the under surface of the leaves, the larvae feed upon the lower epidermis and mesophyll, often leaving the upper skin unbroken, but occasionally the leaves are skeletonised. Pupation takes place in the ground. The whole life cycle occupies 9–10 weeks. There is probably more than one generation per annum.

Genus *Phyllotreta*. Flea Beetles

This genus, like other members of the *Halticinae*, have greatly developed powers of leaping as evidenced by the swollen hind femora.

The beetles belonging to this genus are mostly pests of Cruciferous crops; a number of species deserve consideration and their appearance and life histories are here briefly described; the questions of host plants, damage and control are dealt with collectively. According to Newton (44), from whose work much of this information is derived, it is the following species which are common on cultivated Cruciferae:

Phyllotreta nemorum Linn.   *P. undulata* Kutsch.
P. cruciferae Goeze.   *P. consobrina* Curt.
P. atra Payk.   *P. vittula* Redt.
P. nigripes F.

Phyllotreta nemorum Linn. Turnip Flea; Turnip Fly

DESCRIPTION. *Adult.* Oval, somewhat flattened. Head closely punctured; antennae long, dark, the first three joints much paler. Thorax broad, rounded, closely punctured. Elytra with two broad yellow bands which are slightly waved. Legs, tibiae reddish yellow. Male with fourth and fifth antennal segments swollen. This species is differentiated from *P. undulata* by its larger size, by the pale bases of the antennae and the tibiae which are entirely reddish yellow. Length $2\frac{1}{2}$–$3\frac{1}{2}$ mm. (Fig. 28.)

Egg. Yellow and rounded, surface finely pitted.

Larva. Yellowish in colour with a regular series of blackish, chitinous plates; head almost black; legs and prothoracic shields dark brown; segmental plates brown. Length 6 mm.

LIFE HISTORY. As with all flea beetles, the winter is passed in the adult stage in a variety of places, under bark, in hedgerows, under dead leaves, etc., and often in farm buildings. The beetles emerge in early spring, usually about the middle of April, and, after pairing, lay their eggs. The eggs are deposited usually in the soil and some-times on the epidermis of charlock, radish and other Cruciferous

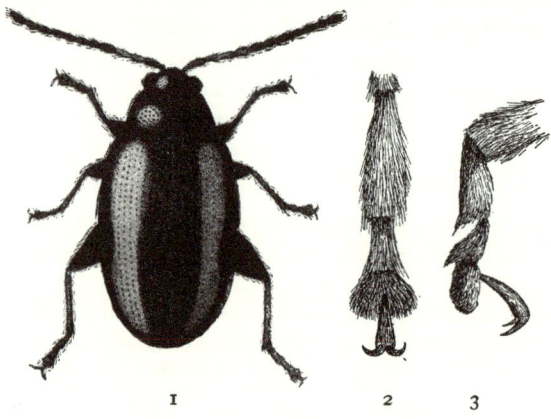

1   2   3

Fig. 28. *Phyllotreta nemorum* Linn. The Turnip Flea Beetle. 1, adult (× 14 approx.); 2, hind tarsus, ventral view, showing the hairy underside (× 75); 3, tarsus, seen from the side, showing the upraised claw (× 75). (After Rostrup.)

plants; they may be laid singly or in batches of 3–4. The larvae hatch in 8–10 days and tunnel in the mesophyll, forming single tunnels at first but later producing bladder-like mines. When full fed the larvae descend into the soil and pupate near the surface in an earthen cell. The larvae may be found in the leaves in the middle of June, the pupae occurring about a month later. The pupal stage lasts about two weeks, the whole life cycle occupying 5–6 weeks. According to Newton (44) there is only one generation a year, but the egg-laying period extends from April to July. In Russia there appear to be from one to six broods per annum. Towards the end of September the adults leave their host plants and hibernate, reappearing in spring and causing the most serious damage.

NATURAL ENEMIES. Newton records the Braconids, *Diospilus*

morosus Reinh., and *Perilitus aethiops* Nees., as parasites of the larvae of *P. nemorum*; *D. morosus* is also recorded from Russia.

Phyllotreta cruciferae Goeze

DESCRIPTION. *Adult.* Deep, shining black with a greenish metallic lustre; antennae with first four joints ferruginous, base testaceous; thorax and elytra almost equally punctured, but the punctures on the latter are arranged in more regular rows. Length 1·8–2·4 mm.

Egg. Pale, yellowish white, elongate-oval, surface finely sculptured, showing polygonal pitting. Length 0·3 mm.

Larva. White, elongate, of narrow diameter; head, thorax and abdomen distinct, three thoracic and nine well-defined abdominal segments; dorsum of first thoracic and ninth abdominal segments slightly more chitinised than the remaining segmental plates. Length 5–6 mm.

LIFE HISTORY. The larvae of this and the remaining species of *Phyllotreta* except *P. vittula*, are root feeders, and the life history here described is typical, with the exception mentioned, of the other species. The eggs are deposited in batches of 20–30 in the soil surrounding the host plant and the larvae feed upon the roots, and when full fed pupate in an earthen cell in the ground. The adult beetles appear in July and August, the maximum emergence being in the first two weeks of August. The total life cycle occupies about six weeks. The Ichneumonid *Gelis* (*Pezomachus*) *carinatus* Först. is recorded from the larva (Newton). This beetle transmits the virus of turnip yellow mosaic (see p. 264).

Phyllotreta atra Payk.

DESCRIPTION. *Adult.* Very similar to *P. cruciferae* but smaller and black in colour without the metallic lustre; legs black, tarsi lighter. Length 1·9–2·5 mm. The larva of *P. atra* is also parasitised by *Gelis carinatus* Först.

Phyllotreta nigripes F.

DESCRIPTION. *Adult.* Rather long and flat, blue or greenish blue, often with a coppery tinge on the thorax; head with six or eight irregular rows of small punctures, antennae black; elytra longer and more finely punctured than in the preceding species; legs black. Length 2–2½ mm. The adults of *P. nigripes* and the other underground-feeding species are parasitised by two Braconids, *Perilitus areolatus* Thoms., and *Mesochorus* sp. (Newton).

Phyllotreta undulata Kutsch.

DESCRIPTION. *Adult.* Oval rather flattened, black; antennae black, bases reddish; thorax broad, densely punctured with a coppery tinge; elytra black with a broad yellow band on each, more or less parallel till the apex where each band curves inwardly; legs, femora black, bases of tibiae brownish, rest of tibiae and tarsi black. Length 2–3 mm. This beetle also transmits the virus of turnip yellow mosaic (see p. 264).

Phyllotreta consobrina Curt.

DESCRIPTION. *Adult.* This species is similar to *P. nigripes* but can be distinguished by its darker colour and stronger punctuation. Male with 4th–5th joints of antennae dilated. Length 2–2½ mm.

Phyllotreta vittula Redt.

DESCRIPTION. *Adult.* A small species, oblong, rather flattened, black with a greenish metallic tinge on the thorax, elytra with a longitudinal yellow stripe on each, which run parallel and are not curved inwards; antennae black, base testaceous; legs black, apex of anterior femora and.under sides of tibiae and tarsi reddish. Length 1½–2 mm. (Fig. 29.)

LIFE HISTORY. The habits of this flea beetle differ from those of the other species of *Phyllotreta* in that the larvae attack Graminaceous plants as well as the Cruciferae. Attacks often occur on barley in April and May, the larvae mining in the stems and occasionally biting through the shoot.

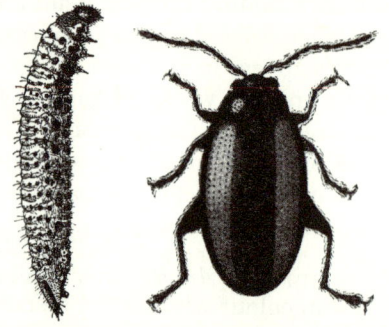

CULTIVATED HOST PLANTS. All the cultivated Cruciferae suffer from the attacks of *Phyllotreta*, the most commonly attacked being white turnip, radish and Brassicas. *P. nemorum* is most attracted to radish.

Fig. 29. *Phyllotreta vittula* Redt. 1, larva (× 8 approx.); 2, adult (× 14). (After Rostrup.)

WILD HOST PLANTS. Among Cruciferous weeds, the following may be mentioned: wild radish (*Raphanus* sp.), *Camelina* sp., treacle mustard (*Erysimum cheiranthoides*), winter cress (*Barbarea vulgaris*), field brassica (*Brassica campestris*), *Sisymbrium* spp., *Thlaspi* spp., *Lepidium* sp. and *Armoracia* sp. Charlock (*Sinapis arvensis*) is often an early host particularly of *P. nemorum*.

SYMPTOMS OF ATTACK AND INJURY TO HOST PLANT. The most serious damage by *Phyllotreta* is done by the adults and there are two periods of attack, the first in the spring by the overwintering beetles and the second about the middle of August by the new generation. It is the spring attack which is of economic importance, when the plants are in the seedling stage. A spell of hot weather at the end of April or beginning of May often brings out the beetles in large numbers and they attack the first two leaves of the plants as they are just appearing through the soil, eating them away and destroying the seedling. According to Newton, there is another and more insidious phase of the spring attack, in which the beetles make their way through the broken surface crust of the soil and bite through the fleshy hypocotyl of the developing seed, thereby destroying the plant. The August attack consists largely of leaf damage, numbers of round holes being eaten in the leaves, but owing to the scarcity of seedling Cruciferae at this period and to the fact that the summer generation tends to remain on the brood plants, this attack is of little significance. The chief features of attack by flea beetles are the suddenness of their appearance, the first day of a hot spell in spring bringing them out in countless numbers, and the rapidity and completeness of the damage owing to the small size of the plant.

DISTRIBUTION. For many years *P. nemorum*, the larger striped flea beetle, has been considered the most common and widely distributed species. Recent investigation, however, indicates that this species is indeed the least common of the *Phyllotreta*. Taylor (56) finds that *P. undulata*, the lesser striped flea beetle, is the common species in the Leeds district as it is also in northern Britain and the Clyde district. In Hertfordshire and Westmorland, *P. undulata* and *P. vittula* are the chief species occurring on turnips; *P. consobrina* is common in parts of Surrey, mostly in sandy and chalky localities.

CONTROL. Temperature plays an important part in the development of flea beetle attack, the first warm day of spring bringing out the hibernating adults in large numbers. Against the early germinating attack below ground, Newton recommends the following measures: (1) rolling to prevent access of the beetles below ground; (2) early sowing to get over the germinating stage before the appearance of the beetles; (3) sowing deterrent dusts with the seed; (4) spraying or dusting with a deterrent as a routine measure. Another method which sometimes gives good results is the steeping of the seed before sowing. Turpentine and paraffin are generally used for this purpose. The conditions necessary for success appear to be that sufficient liquid be used to make the seed thoroughly wet, that drilling should take place as soon as possible after treatment and that germination should not be delayed. (Jenkins (28).) Seed to be treated is spread out on a bag, sprinkled with paraffin and turpentine and then rubbed with the palm of the hand until every seed has received a coating of liquid; about a tea-cup-full of paraffin and rather less of turpentine should be used per pound of seed. As some commercial turpentines are injurious to the seed, it is advisable that a test should first be made by treating a small quantity of seed and germinating it upon flannel or blotting paper. The deterrent effect of this procedure upon the flea beetle is presumably due to the adherence of traces of the turpentine to the young seedling; wet weather soon after germination is therefore likely to be unfavourable to the success of this measure. Against the spring attack on the seedling plants, various other measures are possible; thorough hoeing between the rows is important, the growth of the plants is thereby encouraged and the flea beetles are disturbed. Of deterrent dusts and sprays, derris or naphthaline-silica dusts should now be replaced by D.D.T. or Gammexane. These should be applied on the day that the seedlings appear in numbers, especially if the day is warm and dry. Experiments by Miles (*Rept. Hort. Res. Sta. Bristol*, 1944) showed that D.D.T. (5 per cent.) and Gammexane (P.P. 666) killed the beetles present when they were applied and increased the percentage of plants free from attack. The need for a second dusting must be judged by local and seasonal conditions. Applications of basic slag are recommended to obtain satisfactory establishment of the seedlings.

Plectroscelis (*Chaetocnema*) *concinna* Marsh. Mangold Flea Beetle; Brassy or Tooth-Legged Flea Beetle

DESCRIPTION. *Adult.* Oval, shining, metallic, bronze; under surface black. Head small with a group of large punctures close to each eye; antennae reddish brown with first joint and apical segments blackish. Thorax broad, thickly punctured. Elytra with regular rows of punctures. Legs reddish with tibiae and tarsi darker, posterior femora greatly dilated, dark. There are projections on the hind tibiae. Length $1\frac{1}{2}$–$2\frac{1}{2}$ mm.

Egg. Elongate, oval. Length 0·5 mm.

Larva. White, narrow in proportion to its length, only slightly chitinised, head brown, anal plate light brown with surface roughened. There are three strong setae of approximately equal length on the sternal plate of the prothorax. Length 5–6 mm.

LIFE HISTORY. There appears to be only one generation a year, the adults hibernating in the soil, in tufts of grass or in and around hedgerows. Oviposition commences about the middle of May; the eggs are deposited in the soil surrounding the host plant and the larvae live upon the root system of the host, the life history thus being very similar to certain of the root-feeding species of *Phyllotreta*. Newton (45) considers that this beetle does not attack the hop but is confined in its food plants to mangolds and the Polygonaceae, such as *Rumex* spp., *Polygon* spp. and rhubarb. For attacks upon mangolds, the control measures recommended against flea beetles (*Phyllotreta* spp.) upon Cruciferae may be employed.

Psylliodes chrysocephala Linn. Cabbage Stem Flea beetle

DESCRIPTION. *Adult.* Oval, rather long, greenish. Head reddish; antennae long, lighter at base, 10-jointed instead of 11 as in *Phyllotreta*. Thorax rounded rather broad, very finely punctate. Elytra with regular rows of fine punctures. Legs yellow-brown to red with posterior femora and tibiae darker. Length 3–5 mm. (Fig. 30, 3.)

Larva. Creamy white in colour, head, pronotal shield and ninth

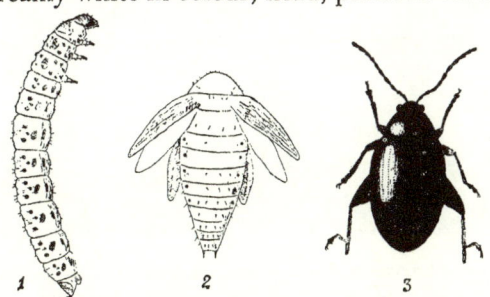

Fig. 30. *Psylliodes chrysocephala* Linn. The Cabbage Stem Flea Beetle. 1, larva; 2, pupa; 3, adult. × 5. (After Rostrup.)

abdominal tergite brown, abdomen with three transverse dorsal rows of hairs. (Fig. 30, 1.) Length 8 mm.

LIFE HISTORY. There are probably two broods of this beetle in the year, but this is not definitely determined; hibernation takes place in the adult stage and probably in the egg and larval stages also. The beetles reappear in early spring and oviposit on or near Cruciferous plants, chiefly Brassicas. The damage to the plant is due to the larvae tunnelling in the stems and midribs of cauliflowers, etc. (Fig. 31); occasionally the stalks are hollowed out, appearing black when viewed externally. The larva, when full

Fig. 31. Larvæ of *Psylliodes chrysocephala* Linn. *in situ* in midrib of cauliflower leaf. Slightly enlarged.

grown, emerges by an exit hole in the stem close to soil level and pupates in an earthen cell in the ground. The writer has found these larvae seriously damaging cauliflowers in Lancashire and Cheshire during May. The larvae, arising from eggs deposited by the overwintering adults, pupate in April and May, the beetles hatching in June, these give rise to a second brood which pass the winter. The adult beetle is not a pest in this country, but is harmful on the continent to the leaves and flowers of rape. Control measures should be directed against the summer brood and consist of collection or trap cropping. Kauffmann (30) gives some account of the parasites of this beetle.

Psylliodes affinis Payk. The Potato Flea Beetle

DESCRIPTION. *Adult.* Oval, ochreous. Head darker; antennae pale, slightly darker at apex. Thorax broad with small punctures and some indeterminate black markings. Elytra with rows of regular rather large punctures which become smaller towards the apex, some black markings anteriorly and on suture. Legs pale, hind femora dark, much dilated. Male with thorax rather narrower. Length 2–3 mm.

Egg. Elongate, oval, yellowish. Length 0·6 mm.

Larva. Whitish, head capsule light brown. According to Newton, the surface of the anal plate is not roughened as in *P. attenuata.* Length 5–6 mm.

LIFE HISTORY. The winter is passed in the adult stage, the beetles reappearing in spring. The eggs are deposited either singly or in clusters in the soil near the host plant; the larvae hatch after about eight days and feed on the roots for about a month, forming galleries. The host plants appear to be Solanaceae, particularly potatoes. There is probably only one generation in the year, the overwintering adults may be found on the wild host plant in May and the new brood in September. The damage caused is due largely to the feeding of the adult which eats out numbers of round holes in the leaves. Like all the potato insect fauna, this insect is under suspicion of being a vector of potato virus diseases (Chapter XIV), otherwise it is of little agricultural importance in this country. In May and September the adults can be found in considerable numbers on the woody nightshade or bittersweet (*Solanum dulcamara*) which is the chief wild host plant. When necessary *P. affinis* can be controlled by the use of Bordeaux mixture and the destruction of the wild host plant.

Psylliodes attenuata Koch. Hop Flea Beetle; Hop Cone Flea

DESCRIPTION. *Adult.* Oval, coppery or greenish. Head small, without punctures; there are two crossed furrows between the eyes;

antennae dark, lighter at base. Thorax thickly punctured. Elytra long with regular and deep punctures. Legs reddish brown, femora darker. Differs from *P. concinna* in the 10-jointed antennae and the hind tibiae. Length 2–3 mm.

Egg. Pale yellow, oval. Length 0·5 mm.

Larva. White, head capsule, prothoracic and anal plates light brown in colour. Length 5–6 mm.

Pupa. White. On the first four abdominal segments are four setae in a cross-line and six on the remaining segments. Length 3 mm.

LIFE HISTORY. The winter is spent in the adult stage in various places, such as the old hop bines, in crevices of hop poles, under the soil or the debris on it; the beetle is also capable of living on nettles till the winter is far advanced, and is often found on these plants in the vicinity of hop gardens. There is one generation in the year, the eggs are deposited in the soil near the hop hills during May and June, the resulting larvae feeding upon the root fibres. The whole life cycle occupies 67–71 days. *P. attenuata* is very injurious to hop gardens in Kent, Sussex, Hampshire and Surrey. The worst damage appears to be done when the hops are in cone, the beetles feeding on the bracts of the cones, riddling them with small holes and so spoiling the market value. No method of control has yet been devised for the attacks on the cones, though numbers of the beetles may be caught on "greased board" traps.

Phaedon cochleariae F. Mustard Beetle

DESCRIPTION. *Adult.* Oval, rather convex, metallic blue, shining; antennae dark. Thorax broad, finely punctured. Elytra, with regular punctured striae, often of a greenish tinge; last abdominal segment reddish at the sides. Legs black. *P. armoraciae* is similar but larger and is less bright in colour. Length 2½–4 mm.

Egg. Ellipsoidal in shape, ochreous yellow in colour, frequently with a slight median longitudinal depression. (Roebuck (49, 50).) Length 1·2–1·5 mm.

Larva. Brownish in colour, hairy, somewhat tapering; head black; prothorax dark; the body is spotted with black and there is a row of brown tubercles along each side from which can be protruded shining yellow glands; legs black and yellow. Length 6 mm.

Pupa. Bright yellow in colour, with a median row of blue spots situate dorsally on the abdomen, clothed with stout brown bristles.

LIFE HISTORY. The adult beetles hibernate in a variety of places such as crevices in palings, in the soil, under loose bark or in the hollow stems of reeds; they are sometimes found in thousands in the stubble of mustard fields. The hibernating beetles leave their winter quarters at the end of April or early in May; the eggs are laid on the leaves of various Cruciferous plants in May and June, each egg being inserted in the tissue in a depression scraped out by the mandibles of the female which may lay 300–400 eggs. The eggs hatch in 8–12 days, the larval stage occupying about 23 days during which three moults occur, the larvae are most numerous during May and June. Pupation takes place in the soil, either with or without an earthen cocoon, at a depth of about 2 in. The whole life cycle from egg to adult occupies about 35 days; there are two overlapping generations in the year. According to Roebuck (50), in the first generation, adults resulting from eggs laid between 15th May and 26th June emerge between 15th June and 30th July, and adults of the second generation from eggs laid between 23rd June and 10th August emerge between 27th July and 10th September.

HOST PLANTS. Mostly Cruciferae, swedes, turnips, cabbages, and particularly mustard; probably many Cruciferous weeds are also attacked. *P. cochleariae* is occasionally a serious pest in watercress beds (*Nasturtium officinale*). The most serious damage is done to the seed crops, the adults frequently attacking the developing seed pods. The chief wild host plant is the large bitter cress (*Cardamine amara* Linn.); also charlock (*Sinapis arvensis*), and wild radish (*Raphanus raphanistrum*).

DISTRIBUTION. Common and fairly widely distributed throughout the United Kingdom; *P. armoraciae* is scarcer in the north of England. In 1924 large numbers of *P. cochleariae* appeared south of Wainfleet in Lincolnshire, and in 1927 serious damage was caused by this insect to cabbages in Kent, to white mustard and cauliflowers in Lincolnshire and to watercress in the Midlands and Wiltshire. Another species, *P. tumidulus* Germ., is occasionally a pest of celery, notches being eaten out of the edges of the leaves; the heart of the plant is also attacked. This beetle has also been recorded from parsley.

CONTROL. These measures include the removal so far as possible of Cruciferous weeds, especially hollow-stemmed ones; mustard stubble should be burned by the middle of October to destroy the hibernating adults. A lead arsenate spray has given good results, while dusting with calcium arsenate has also been recommended.

NATURAL ENEMIES. Miles (40, 41) records the Tachinid fly *Meigenia floralis* Fln. as a parasite of both species of mustard beetles, and Baer (1) the Tachinid *Morinia pullula* Zett. from Europe. Falcoz (13) in France records the Histerid beetle *Saprinus virescens* Payk. as predacious upon the larva of *P. cochleariae*.

Cassida vittata Vill. Small Green Tortoise Beetle

This insect is seldom a serious pest and is dealt with very shortly. It is a small green species, the head hidden under the pronotum, and the elytra rounded at the extremity, broader than the abdomen with a raised lateral margin, which gives the insect a flattened shield-like appearance. The larva is somewhat flattened and shiny with a forked caudal process; it is covered with a layer of excrement and cast skins which masks its true appearance. *C. vittata* is occasionally a pest of mangolds and sugar-beet, and an outbreak was recorded in Sussex in 1922. The goosefoot (*Chenopodium* spp.) is probably the wild host plant for this beetle.

NATURAL ENEMIES. In Italy the larvae of *C. vittata* are parasitised by *Tetrastichus bruzzonis* Masi, *Brachymeria vitripennis* Först. and *Habrocytus* sp. *Trichogramma evanescens* Westw. was bred from the eggs. (See Menozzi, C., *Indust. saccarif. ital.* XXIII, 1, 2, 4, Genoa, 1930.)

Family *Nitidulidae*

Small flat insects, antennae usually with eleven joints, the last three forming a club; coxae cylindrical, tarsi 5-jointed.

Meligethes aeneus F. Pollen Beetle; Blossom Beetle

DESCRIPTION. *Adult*. Oval, with a greenish bronze tint, finely punctured, and a fine pale pubescence. Head small, triangular; antennae dark with the first joint thickened, second reddish brown, last three joints form a round club. Thorax rounded, broad, rather shining. Elytra with very fine reticulate markings. Legs dark, rather shiny; femora and tibiae flattened and dilated, outer margin of anterior femora finely toothed. (Fig. 32.) Length $1-2\frac{1}{2}$ mm.

Larva. Whitish yellow in colour; head black, somewhat bristly; behind the head is a sooty patch; body tapering somewhat, posteriorly; an anal pro-leg is present; legs whitish, black at tips.

LIFE HISTORY. The importance of this insect is mainly due to its connection with seed crops. The overwintering adults appear in the spring and oviposit in the flowers of Cruciferous plants such as mustard, rape, swedes, turnips, etc. The chief wild host plant is Charlock (*Sinapis arvensis*). The eggs are white, elliptical in shape and about ½ mm. long, they hatch in 7–10 days (in Russia, the period is given as 4 days). Comparatively little is known of the biology of the larvae, which live concealed in the buds and flowers and feed largely on the pollen. The larvae are only likely to develop into pests when they are present in abnormal numbers, and if the infesta- tion does not exceed 25 per cent. of the buds and flowers they may even have a beneficial effect as pollen carriers. There is only one larval moult, the larvae pupating in the second instar in the soil in earthen cells close to the host plant. The length of the larval stage is about 13 days and that of the pupal stage about 16 days. There is probably more than one generation in the year, the adults of the last brood hibernating in the soil. The most serious damage is done by the adult beetle and becomes apparent when the plant begins to flower. If the beetle appears when the plants are in full bloom, it often eats the nectaries when feeding on the pollen and sucking the nectar, but no further injury is done. If, how- ever, it appears early before flowering time, it makes its way through the bud in search of pollen and destroys it. The beetle is one of the commonest found living on flowers and is sometimes a serious pest in districts where Cruciferous crops are grown for seed. It is a troublesome pest in Germany. Control measures should include early planting and the selection of early flowering varieties of Crucifers to avoid bud injury and the preparation of a good seed bed. A lead arsenate dust has also been used with some

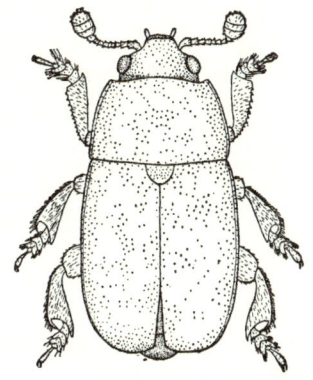

Fig. 32. *Meligethes aeneus* F. Pollen Beetle. × 22. (After Imms.)

success, or spraying with 3 lb. Paris green, 5 lb. unslaked lime to 60 gallons of water with starch paste to ensure adherence is a useful measure. Probably the best method of control is now a dust of D.D.T. or Gammexane.

NATURAL ENEMIES. The Ichneumonids *Thersilochus morionellus* Holmgr. and *Isurgus heterocerus* and the Braconids *Diospilus oleraceus* Hal. and *Microgaster* sp. are recorded in Europe. The beetle *Coccinella septempunctata* is predacious on the larva.

Family *Hydrophilidae*

Mainly aquatic or sub-aquatic beetles; a considerable number, however, are inhabitants of marshy places, and certain members of the genus *Helophorus* attack turnips and are briefly dealt with.

Genus *Helophorus*

The beetles of this genus secrete a substance on the surface of the body to which small particles of soil adhere, covering the insect with an incrustation and thereby rendering description difficult. *H. rugosus* and the closely allied *H. porculus* have been studied by Petherbridge (47), from whose work some of this information, including the descriptions of the stages, is derived.

Helophorus rugosus Ol. Turnip Mud Beetle

DESCRIPTION. *Adult.* Head dark, reddish, flattened; palps as long as antennae, both brown; the antennae are pubescent, 9-jointed, the last four segments forming a club. Thorax as broad as elytra, strongly ridged. The pronotum is large and strongly arched, the raised median portion forms a hood over the head, this portion is also marked with three grooves, having four warty prominences between. Elytra reddish brown with indeterminate black markings, longitudinally ridged with coarse punctures in between; forms with the elytra darker sometimes occur. Legs reddish brown. Length 4–6 mm.

H. porculus similar. Fowler (15) differentiates the species as follows: "it may be distinguished from *H. rugosus* by its average smaller size, flattened dorsal costae of thorax, and the fact that the elytra are not sinuate near the base, and have the humeral angles rounded, whereas in *H. rugosus* the elytra are sinuate before the base and the humeral angle is turned outwards forming a distinct tooth". Length 3·5–4·5 mm.

Larva. Cream coloured, cylindrical, tapering, soft and fleshy, with characteristic brown chitinous plates situate dorsally on the thoracic and abdominal segments. The head is narrower than the

COLEOPTERA 129

prothorax, and the antennae are 3-jointed and yellowish brown in colour. The mandibles are stout and brown in colour, each bearing three blunt teeth; the legs are short, terminating in a simple claw, the abdomen ends in a pair of anal cerci. The larva of *H. porculus* is similar. Length 10 mm.

Pupa. Soft and whitish, bearing a number of long bristles arising from prominent conical tubercles; the head is bent beneath the prothorax and bears three pairs of long bristles; the ninth abdominal segment terminates in a pair of pointed lobes. Length 6 mm.

LIFE HISTORY AND CONTROL. The winter is passed in the larval stage and the adults also are capable of hibernating. The eggs are deposited in July and August and the resulting larvae feed through the winter on turnips until the following March or even later. When full grown the larva hollows out a cell, 2 in. below the soil level, in which it pupates, the adult beetle emerging about three weeks later. The host plants are mostly Cruciferae, especially the white turnip (*B. rapa*); other hosts include swedes, rape and occasionally lettuce. Both larval and adult stages cause injury, the most serious being the occasional destruction by the larvae of the growing point of young turnips. Other damage consists of tunnelling or boring in the bulb of the turnip; the leaves are also attacked at the edges by both stages. Control is restricted to cultural measures; turnips should be sown in good time where *H. rugosus* is troublesome; as the larvae feed in the fall of the year, a late summer fallow is recommended to starve any larvae present. It is inadvisable to grow turnips on land which bore a crop attacked by the beetle in the preceding year.

Family *Silphidae*. Burying and Carrion Beetles

Blitophaga (*Silpha*) *opaca* Linn. Beet Carrion Beetle

DESCRIPTION. *Adult.* (Fig. 33.) Oval, rather flattened, clothed with dark orange to reddish pubescence which gives the insect a bronze tint. Head rather square behind eyes; a number of strong reddish bristles at the base of the antennae which are rather long with the last three segments dilated. Thorax broad, hood-like, strongly punctate, anterior margin slightly, posterior margin strongly, sinuate; there are two smooth patches devoid of punctures in the centre of the dorsum. Elytra black, each with three raised ridges, the outermost one most pronounced and ending in a raised tuberculate area. Legs black with some paler spines. Fowler gives the first four joints of the anterior tarsi as dilated in the male. Length 10 mm.

Larva. Broad, flattened, black and shining, with the edges of the abdominal tergites backwardly directed, the last abdominal segment bears a pair of processes; the appearance of these larvae is very suggestive of a woodlouse. (Fig. 33.) Length 10–12 mm.

LIFE HISTORY. The Silphidae are mostly carrion feeders, but one or two species, of which this is an example, have adopted the phytophagous habit. The adult beetles pass the winter under stones and

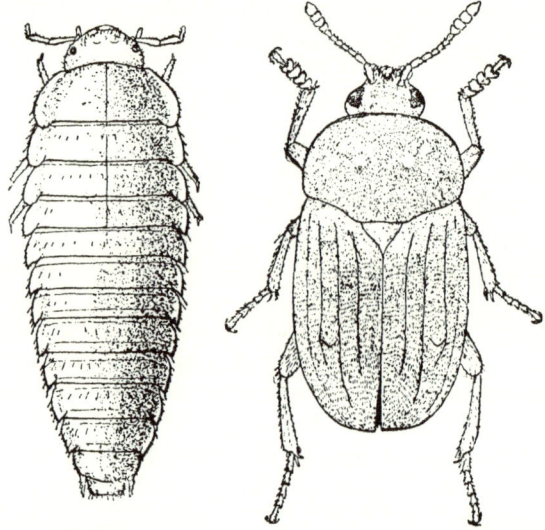

Fig. 33. Larva and adult of *Blitophaga opaca* Linn. The Beet Carrion Beetle. Much enlarged. (From Imms after Kemner.)

moss, reappearing in May and early June when they migrate to young beet and mangold plants. The eggs are deposited in the soil; after 5–9 days the larvae hatch and make their way to the young leaves where they feed for about three weeks; after which they descend into the soil and pupate about 1 in. below the surface in an earthen cocoon. Both adults and larvae feed upon the leaves, the attack beginning at the edges which frequently turn black, later the larvae may turn their attention to the roots which they eat through just below ground level. On the continent, particularly in Germany, this insect may be a severe pest; according to one writer the progeny of one pair of beetles are capable of defoliating 100 square yards of cultivated land. In the British Isles it is not usually a serious pest, but an account is given of it as being a potential insect enemy of sugar-beet.

CONTROL. Control measures consist in frequent hoeing to disturb the pupae and also the eggs which are susceptible to drought; destruction of the wild host plants such as goosefoot (*Chenopodium* spp.) when practicable is advisable. The following sprays are recommended: Paris green at the rate of 3–4 lb. per 100 gallons of water, to which has been added 3–4 lb. of chalk and $\frac{1}{5}$ lb. of gelatine to make it adhesive, or, alternatively, 50 lb. barium chloride and 20 lb. wheat flour to 100 gallons of water.

Family *Bruchidae*

Head shortly and flatly produced somewhat after the manner of weevils with which they are sometimes confused. Elytra shortened leaving the tip of the abdomen exposed. The larvae live mostly in the seeds of Leguminosae.

There are three species of the genus *Bruchus* associated with peas and beans in the British Isles; they are: *Bruchus pisi* Linn., *B. rufimanus* Boh., and *B. affinis* Frol. These beetles are known variously as Bean Beetles, Broad Bean Beetles and, erroneously, as Pea and Bean Weevils (p. 135). Their life histories and habits are similar.

Bruchus rufimanus Boh. Bean Beetle; Broad Bean Beetle

DESCRIPTION. *Adult.* Oblong, rather flattened, black; first four joints of antennae reddish and smaller than the remaining joints which are darker. Thorax broad, rounded and narrowing slightly in front, covered with a reddish pubescence. Elytra black with regular striations, patches of whitish pubescence occur here and there giving a mottled appearance, there is also usually a stripe of white pubescence along the suture; pygidium covered with a silvery pubescence. Legs black, anterior femora red. Length 3–4 mm. (Fig. 34.)

Bruchus pisi Linn. Pea Beetle

Adult. Very similar to the foregoing; there is usually more white on the elytra and the anterior femora are black; pygidium with two bare black markings. Length 3–4 mm.

Bruchus affinis Frol. The Spanish Bean Beetle

Adult. Similar to the other two species but distinguished by the strong and projecting teeth on the thorax and by the finer sculpturing of the thorax and elytra.

Egg. Elongate, oval, white or greenish yellow.

Larva. Soft and fleshy, white or cream coloured; legs usually absent or they degenerate later; strong mandibles; head scaly. A number of spines are present on the first abdominal segment which are thought to be of use when boring in the bean or pea pod. Length 5–6 mm.

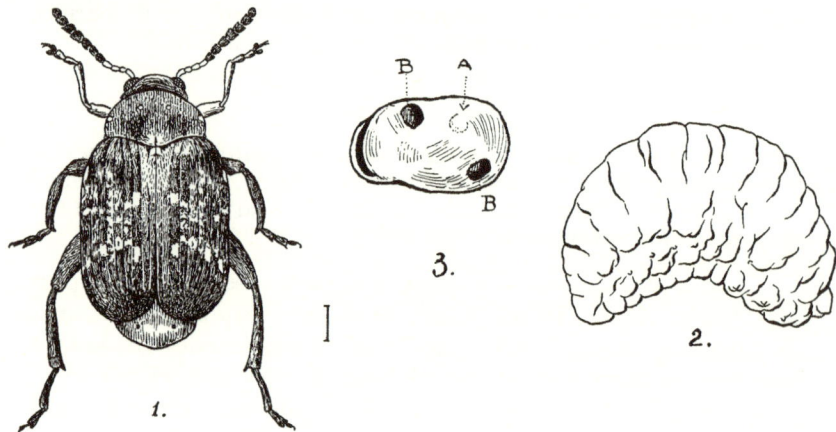

Fig. 34. *Bruchus rufimanus* Boh. The Common Bean Beetle. 1, adult beetle; 2, larva; 3, attacked bean. *A*, hole covered by skin, insect inside. *B*, empty holes from which beetles have emerged. (Reproduced by permission of the Ministry of Agriculture.)

LIFE HISTORY. The overwintering beetles appear in spring, sometimes occurring on gorse bushes as early as March, others remain in the bean or pea until May. The female deposits her eggs on the young seed vessel in the pea or bean blossom or on the outside of the pod after the blossom has fallen, the eggs hatching in about 12–15 days. The young larvae leave the egg by the side attached to the pod and bore their way into the young beans or peas; in the case of *B. pisi* only one larva enters the pea. The larva eats out a cell in the bean, its position being indicated by a transparent spot. Pupation takes place in the seed, either when ripening or after harvest, the adults hibernating in the seed. It is a moot point as to whether the broad bean is very seriously injured by the larvae of *B. rufimanus* which as a rule appear to avoid the radicle and plumule. Larson (33), however, considers that the planting of infested beans reduces the yield as the injury may prevent germination, accelerate decomposition of the germinating seeds or prevent the cotyledon from developing properly. Infestation, however, does not appear to have any bearing on the infection of the succeeding crop.

DISTRIBUTION. *B. rufimanus*, which is common and widely distributed, is probably the only indigenous species of the three considered here. *B. affinis* is annually introduced from the Mediterranean countries on such bean seeds as Seville Longpod and does not seem able to establish itself permanently. *B. pisi* is also continually introduced with imported peas and it is doubtful if it could maintain itself in the British Isles without reintroduction on seed peas from warmer climates.

CONTROL. Both adults and larvae are easily killed in the dried seed by fumigation and this should always be practised with infested seed. Either carbon bisulphide or sodium cyanide may be used; with the former use 3 lb. carbon bisulphide per 1000 cubic feet at a temperature of 58° F. for 24 hours or half the quantity for 48 hours; with sodium cyanide use 3–4 lb. at 70° F. for 48 hours; an alternative substance is carbon tetrachloride used at the rate of ¼ lb. per 50 cubic feet. This has the advantage of being non-inflammable. Fumigate in an air-tight container with caution owing to the highly poisonous nature of the fumigants.

NATURAL ENEMIES. Two Hymenopterous parasites are recorded from *B. rufimanus* in Britain, a Pteromalid (Chalcididae), *Bruchobius laticeps* Ashm. and a Braconid, *Sigalphus luteipes* Thoms. *B. laticeps* is also recorded from *Bruchus pisi* in France, and *Sigalphus primus* Brethés (n.sp.) from *B. rufimanus* in the Argentine.

Family *Curculionidae*. Weevils

Head drawn out into a rostrum; antennae usually geniculate and clubbed. The larvae are legless and exhibit great similarity of form; they are curved, fleshy grubs; the head is rounded and yellowish, the rest of the body soft, whitish and furrowed. They are mostly internal feeders and occur in all parts of the plant host.

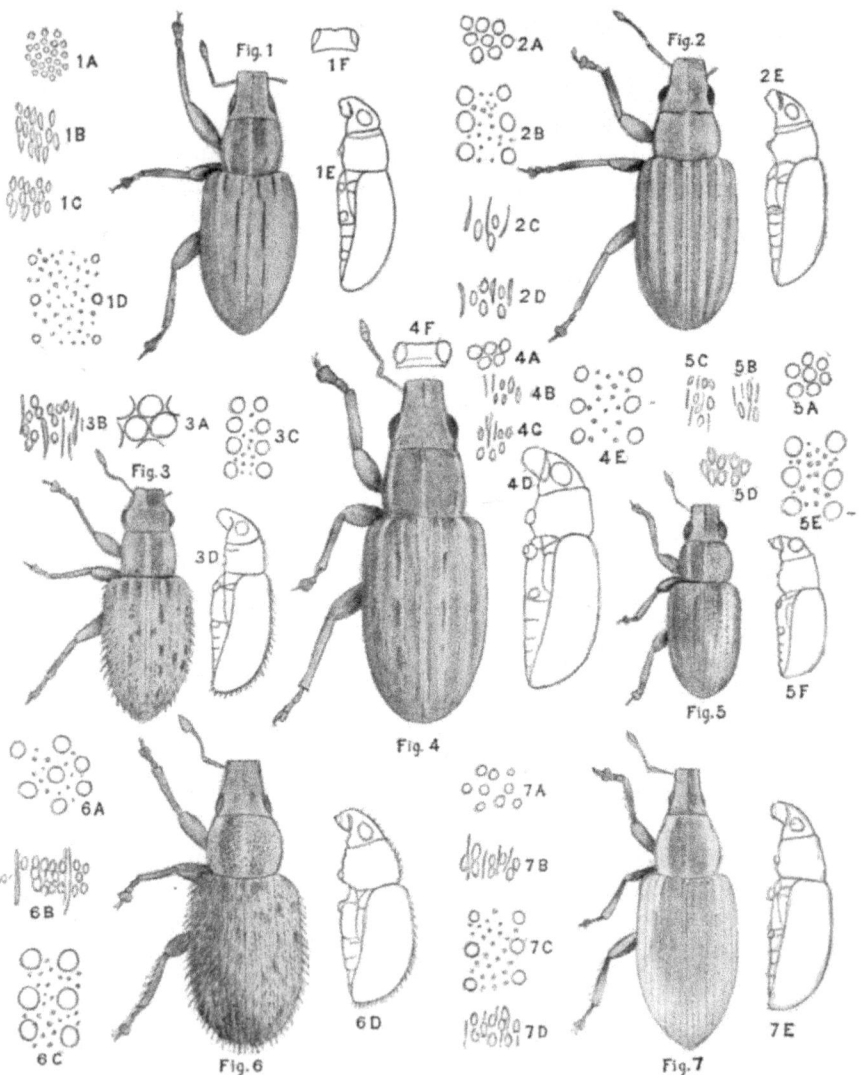

Fig. 35. *See opposite page.*

Genus *Sitona*

A number of weevils belonging to this genus are serious pests of clover and pulse crops; the following species habitually attack these crops:

Sitona lineata Linn.   *S. humeralis* Steph.
S. hispidula F.   *S. flavescens* Marsh,
S. crinita Herbst.   *S. sulcifrons* Thunb.
S. puncticollis Steph.

Of these *S. lineata* is the commonest and best known; it is often a serious pest of peas. Much of the following information is derived from the work of Jackson (25), who has made a study of the group.

Sitona lineata Linn. Pea Weevil

DESCRIPTION. *Adult.* Black, clothed with fuscous scales which give the insect an ochreous appearance. Head, rostrum with a central furrow which is continued between the eyes which are prominent; antennae rather long, light brown. Thorax slightly broader than head with three light lines. Elytra with parallel lines alternately light and dark and fine punctured striae, the lineations vary greatly in intensity and may be almost absent. Legs rather long, testaceous, femora darker, anterior tibiae curved in the male. Length 4–5 mm. (Fig. 35, 2.)

Egg. Oblong oval to round, whitish in colour when first laid, later changing to black, surface slightly roughened. Length 0·36 mm.

Fig. 35. Weevils of the genus *Sitona* injurious to leguminous crops in Britain. × 7 approx. 1. *S. flavescens* Marsh. 1 A, punctuation of thorax; 1 B, scales of thorax; 1 C, scales of elytra; 1 D, punctuation of elytra; 1 E, side view of weevil; 1 F, forehead viewed from above. 2. *S. lineata* Linn. 2 A, punctuation of thorax; 2 B, punctuation of elytra; 2 C, scales and setae of thorax; 2 D, scales and setae of elytra; 2 E, side view of weevil. 3. *S. crinita* Herbst. 3 A, punctuation of thorax; 3 B,. scales and setae of elytra; 3 C, punctuation of elytra; 3 D, side view of weevil. 4. *S. puncticollis* Steph. 4 A, punctuation of thorax; 4 B, scales and setae of thorax; 4 C, scales and setae of elytra; 4 D, side view of weevil; 4 E, punctuation of elytra; 4 F, forehead viewed from above. 5. *S. sulcifrons* Thunb. 5 A, punctuation of thorax; 5 B, scales and setae of thorax; 5 C, scales and setae of elytra; 5 D, scales from sides of body; 5 E, punctuation of elytra; 5 F, side view of weevil. 6. *S. hispidula* F. 6 A, punctuation of thorax; 6 B, scales and setae of elytra; 6 C, punctuation of elytra; 6 D, side view of weevil. 7. *S. humeralis* Steph. 7 A, punctuation of thorax; 7 B, scales and setae of thorax; 7 C, punctuation of elytra; 7 D, scales and setae of elytra; 7 E, side view of weevil. (After D. J. Jackson.)

136   COLEOPTERA

Larva. Cylindrical, curved, tapering slightly towards the extremities, creamy white and wrinkled, legless; the tenth abdominal segment forms a fleshy prominence which is used in walking, it can be extruded or withdrawn; head small, eyes absent. Length 4–5 mm.

LIFE HISTORY. The adult beetles hibernate in a variety of places, such as long grass, stacks of pea straw, stubble of clover fields, in the soil and often in and about farm buildings. In the first warm days of spring the beetles reappear and the females deposit their eggs in moist earth near the bases of the pea plants. The larvae hatch in about 21 days and make their way into the root nodules in which they burrow. The length of the larval life is 6–7 weeks; when full fed, the larva pupates in a cell about 2 in. below the soil level, the adult emerging about 17 days later. These beetles appear during July and August and are the hibernating adults, they do not become sexually mature until the following spring but feed for a time upon peas and beans; later, when the crop is harvested, they migrate to clover and lucerne. A few weevils may remain on the clover throughout the summer; there is only one generation in the year.

HOST PLANTS. Peas, beans, lucerne (*Medicago sativa*), black medick (*M. lupulina*), all species of clover (*Trifolium* spp.), and vetches (*Vicia* spp.). Abroad it has been recorded as damaging sugar-beet, raspberries, chicory (*Cichorium*), vines, and roses.

INJURY TO HOST PLANT AND SYMPTOMS OF ATTACK. Both adults and larvae are injurious, the adults after hibernation doing most damage. They attack the plants in spring when about 3–6 in. high, feeding upon the leaves in a characteristic way. Commencing at the edges of the leaves U-shaped notches are eaten out and the same damage is done to the unopened shoots. In cold springs, when growth is checked, the whole of the young leaves and growing shoots may be eaten away, but injury may also be marked when the early development of the peas occurs in warm dry weather. The larvae attack the root nodules and hollow them out, the maximum destruction developing at the commencement of the flowering season.

DISTRIBUTION. Very common and widely distributed; severe attacks on peas were noted early in the season in 1925 in Cambridgeshire and the south-eastern counties, mid-Wales and to a

COLEOPTERA 137

lesser extent in the east midlands. The beetle was plentiful in Devon in 1926 and in Rutland and North Wales in 1927.

NATURAL ENEMIES. Jackson (25) records the Braconids *Perilitus rutilus* Nees., *Pygostolus falcatus* Nees. and *Leiophron muricatus* Hal. var. *nigra* from the adult beetle. The fungus *Botrytis bassiana* is also an effective parasite. In Russia the Braconid *Perilitus labilis* Ruthe is recorded.

Three of the other species of *Sitona* mentioned are pests of clover and lucerne and are dealt with briefly. They are *S. hispidula*, *S. sulcifrons* and *S. crinita*.

Sitona hispidula F.

DESCRIPTION. *Adult.* A much darker species than *S. lineata*. Head short and with eyes less prominent. Thorax with sides rounded, heavily punctured, with three lighter lines. Elytra black with some fuscous scales, regular rows of deep punctures and stiff, scattered grey bristles. Legs, femora black, tibiae and tarsi reddish. Length 3–4 mm. (Fig. 35, 6.)

Larva. Very similar to that of *S. lineata* but is more translucent and greyish in colour.

LIFE HISTORY. The adults live for about 12 months, hatching from July till September, the habits differ from those of *S. lineata* in that egg-laying commences 6–8 weeks after emergence and may continue during the winter; hibernation therefore occurs in both adult and egg stage. There is only one generation a year and the insects are more or less confined to clover. Much damage may be done by the larvae, which feed upon the roots of the plants.

NATURAL ENEMIES. In Britain, the Braconids *Perilitus aethiops* Nees., *P. rutilus* and *Pygostolus falcatus* Nees. are recorded, and the Scelionid *Anaphes* sp. is found in Russia.

Sitona crinita Herbst.

DESCRIPTION. *Adult.* Reddish brown to black clothed with greyish or yellowish scales; antennae rather short and reddish, club dark. Thorax with black markings in the form of two parallel bands. Elytra with a number of black patches from centre to apex and regular rows of rather sparsely set setae. Legs, femora dark, tibiae and tarsi of a reddish tint. Length 3–4 mm. (Fig. 35, 3.)

LIFE HISTORY. Very similar in its habits to *S. lineata* and occurs with it on the plants. It is said to occur more frequently on the upper parts of the plant as compared with the lower parts preferred by *S. lineata*. The young adults often attack wild perennial leguminous plants.

NATURAL ENEMIES. The Scelionid *Anaphes* sp. is recorded from Russia.

Sitona sulcifrons Thunb.

DESCRIPTION. *Adult.* Black, elytra with copper coloured scales, smaller than *S. lineata*. The habits and life history of this species are very similar to those of *S. hispidula*. Length 2·9–4 mm.

NATURAL ENEMIES. The Braconid *Perilitus cerealium* Hal. is recorded from *S. sulcifrons*.

CONTROL. No very efficient methods of control are known against *Sitona* spp. Dusting with a powder containing pyridine has been recommended or a 1 per cent. mixture of chlorcresylic acid and precipitated chalk at the rate of 2 oz. per square yard; the following sprays have also been used with some success: Paris green (1 lb. to 120 gallons of water), a 4 per cent. solution of barium chloride, or lead arsenate. A wash consisting of 2 oz. of washing soda and 1 oz. of carbolic soft soap to 1 gallon of water applied to the ground round the plants has proved of use in some cases. Other measures include dusting with soot and heavy rolling while the plants are about 4 in. high. It is important to encourage rapid growth of the plants and clear away all weeds and rubbish. Perennial and annual leguminous crops should not be planted in close proximity.

Ceuthorrhynchus pleurostigma Marsh. Turnip and Cabbage Gall Weevil

DESCRIPTION. *Adult.* (Fig. 36.) Black, shining, body rather stout, with scanty white pubescence, ventrally with whitish scales; antennae inserted a little in front of the middle of the proboscis, geniculate, 12-jointed with the basal joint long and clubbed; rostrum long and curved. Thorax rather large, narrowing at apex, anterior margin, raised and collar-like, coarsely punctured, with a central groove and a small tubercle on each side. Elytra with ten finely punctured striae on each. Legs black with silvery pubescence; femora with a small tooth on the under side. Length 2–2½ mm.

Egg. Very small, transparent, oval or roundish, soft. Length 0·35 mm.

Larva. This is a typical weevil larva, curved and legless, whitish, rather transparent; head small but distinct, body wrinkled

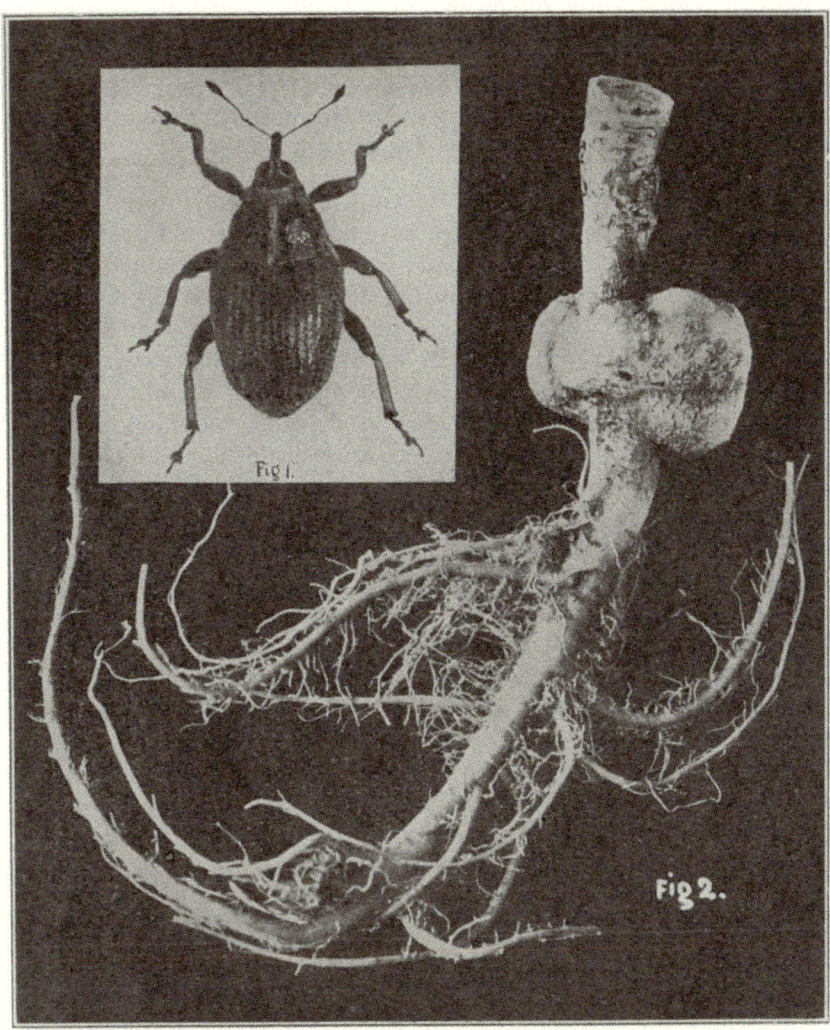

Fig. 36. *Ceuthorrhynchus pleurostigma* Marsh. The Turnip Gall Weevil. Greatly enlarged. 2, cabbage root showing galls. (Reproduced by permission of the Ministry of Agriculture.)

transversely; there is a lateral row of tubercles, one on each segment, the largest at the hinder end of the body, dwindling in size towards the head. Length 4 mm.

Pupa. Free, soft, whitish; first two pairs of legs folded over the wing sheaths, third pair beneath the wings. On the surface of the body are numerous tubercles, bearing protective spines, the abdomen ends in a pair of hooks. Length 3·7 mm.

LIFE HISTORY. According to Isaac(24), there are two separate races of this insect; these are not successive broods but are independent of one another. One, the spring race, lives upon charlock and does not attack cultivated Cruciferae at all; the other, the second or summer race, lives upon cabbage, turnips, etc., and is of some agricultural importance. As regards the spring race, the beetles originate from eggs laid in early spring by adults of the previous year's spring race; these adults do not become sexually mature in the year of hatching. Hibernation of this race therefore is in the adult stage. Egg-laying takes place from March to May, the adults appearing at the end of July and in August; breeding is confined to the roots of charlock. In the summer race, hibernation takes place in the larval stage in the gall on the root of the host plant. The eggs are laid about the end of August in the roots of cultivated Cruciferae and it is the resulting larva which hibernates. The eggs are deposited singly, each in a hole made in the bark of the root by the rostrum of the female; the eggs hatch in 5–17 days according to the temperature, and the larva feeds throughout the autumn, hibernating, when the cold weather comes, in the gall which is formed about it by the plant. The larvae recommence feeding in spring; they leave the root of the host plant in March and April and pupate in the soil, the adults emerging in May and June.

HOST PLANTS. The spring race apparently attacks only charlock (*Sinapis arvensis*), but the summer race, which alone is of economic importance, attacks all types of cultivated Cruciferae, particularly turnips and cabbage seedlings.

INJURY to HOST PLANT AND SYMPTOMS OF ATTACK. Attack by the larvae of *C. pleurostigma* can easily be recognised by the characteristic marble-like galls or swellings on the roots and underground portions of the plant which are set up in response to the irritation caused by the presence of the larva. (Fig. 36, 2.) These galls externally

COLEOPTERA 141

resemble somewhat the swellings due to the 'finger and toe' fungus (*Plasmodiophora brassicae*), but are at once distinguished from them by their round shape and by the presence of the larva or its empty cell and exit hole. It has been a matter for discussion as to whether the galls seriously interfere with the growth of the plants; the writer has noticed, however, that in setting out, cabbage seedlings so affected often fail to take, especially during a spell of hot weather. Reports during 1926–7 also suggest that the insect may be a serious menace, especially in the broccoli-growing area of west Cornwall.

DISTRIBUTION. Widely distributed over the British Isles and Europe generally. Reports to the Ministry of Agriculture during the last few years show it to be plentiful in west Cornwall, the Bristol district and near Hereford.

CONTROL. Infected plants which have overwintered should be burned before March; ploughing the land deeply when the infested crop has been removed is recommended. If infested plants are stacked in such a manner that they dry rapidly, the larvae are unable to emerge and so die. (Isaac (24).) Crop rotation should be practised, bearing in mind that only crops grown from July onwards are liable to infestation. In seed beds, oviposition may be partly prevented by dusting with a mixture of one part sulphur, two parts soot and two parts gypsum, or one part of gypsum containing 2 per cent. of creosote.

NATURAL ENEMIES. Slugs show a fondness for the galls and, in eating these, frequently destroy the larvae. The larvae of the turnip mud beetle (*Helophorus rugosus*) have also been recorded as preying upon the weevil larvae in the galls. The Braconid *Diospilus oleraceus* Hal. parasitises the summer race. The Braconid *Perilitus melanoplus* Ruthe and the Ichneumonid *Thersilochus truncorum* are recorded from *C. pleurostigma* in Europe.

Ceuthorrhynchus assimilis Payk. Turnip Seed Weevil; Cabbage Shoot Weevil

DESCRIPTION. *Adult.* Very similar to *C. pleurostigma*, but can be distinguished by its leaden colour due to the greyish white scales on the elytra, which are more heavily punctured and a little narrower with the shoulders more prominent; the tooth on the femora of *C. pleurostigma* is absent in this species. Length 2–3 mm.

The egg and larval stages are similar in appearance to those of *C. pleurostigma.*

LIFE HISTORY. The overwintering adults appear in the spring and lay their eggs in the seed vessels of Cruciferous plants during May. The female bores a hole in the pod with her rostrum and deposits inside from one to three eggs, close to the ripening seeds. The larvae feed upon the pistils of the flowers and the unripe seeds, often destroying all the seeds in the pod, which becomes discoloured and sometimes deformed. The attacked pod opens prematurely causing the larvae to fall to the ground, where they pupate at a depth of 1–2 in. in earthen cells. The pupal period lasts for 2–4 weeks and there are probably two broods in the year.

HOST PLANTS. These include most of the cultivated Cruciferae and some wild species such as charlock (*Sinapis arvensis*). In Germany *C. assimilis* is a serious pest of rape.

CONTROL. Little is known of control measures for this beetle; arsenical sprays have proved ineffective. The presence of Cruciferous weeds and clay soils are factors favourable to the insect. After an attack on mustard or rape, deep ploughing of the stubble is recommended. Theobald advises mechanical collection of the adults. Turnip seed should be examined as it frequently contains pupae; if these are present the seed should be fumigated.

NATURAL ENEMIES. The Pteromalid *Trichomalus fasciatus* Thoms., the Braconid *Perilitus melanoplus* Ruthe and the Proctotrypid *Platygaster* sp. are recorded from *C. assimilis* in Germany.

Cryptorrhynchus lapathi Linn. Willow Weevil

DESCRIPTION. *Adult.* A rather large black and yellowish weevil; rostrum long and curved, held beneath thorax when at rest; antennae reddish with an oval club. Thorax narrowing towards apex, sides rounded, dorsum with numbers of flattened, yellowish white scales; anteriorly are a number of excrescences formed by flattened black scales; scutellum round, black, finely punctate. Elytra with white scales at base; there are two raised tubercles of black scales on each elytron; punctures of striae in the form of large indentations; apex of elytra with white scales. Legs black, thickly set with white scales; femora with two small teeth. Length 8–9 mm.

Larva. This is of the usual weevil type, whitish, wrinkled and fleshy. Length 6–8 mm.

LIFE HISTORY. The eggs are deposited singly in cavities in the stumps, or on 2–3 year old nursery plants in the corky portion of the tree. The eggs also are often laid near buds or in malformations caused by pruning. The young larvae hatch in 18–20 days, and bore in the stumps or mine in the rods of two years' growth, boring in the heartwood or sapwood of the host. In the British Isles the larvae pupate in chambers at the end of the burrow, the adults hatch in the autumn but remain in the burrow till the following spring. Probably some larvae also hibernate as such; the life history of this insect seems to differ in some respects on the continent. In Holland and Germany eggs laid in the summer do not hatch until the following March. This weevil is sometimes a severe pest and is common in all willow-growing areas, notably Somersetshire. Much injury is caused by the larvae mining in the stumps, and in the young rods, which readily break in the wind; the adults also do considerable damage by gnawing the bark and sapwood of the rods. Affected trees can be recognised by the borings and wood debris from them, surrounding the stools. Control measures consist in digging out and burning infested stools and the collection of the adults by shaking them on to sheets. Matheson (37) recommends spraying the trees in March before the foliage has appeared, or in autumn after the fall of the leaf, with a carbolineum spray made up as follows: 1 lb. sodium carbonate, 1 quart of hot water, and 1 quart of carbolineum.

Apion apricans Herbst. (*assimile* Kirby). Clover Weevil

DESCRIPTION. *Adult.* A very small pear-shaped insect, black; antennae black, base reddish-yellow; rostrum long, rather curved and shining. Head rugosely punctured with a furrow between the eyes. Thorax rather broader than head, regularly and coarsely punctured. Elytra coarsely punctured with regular striae. Legs black; femora and anterior tibiae reddish. Length 2 mm.

Larva. This is of the usual weevil type, very small, dirty white to dusky with a reddish head.

LIFE HISTORY. The adults hibernate in sheltered places, in hedgerows, etc., emerging from about the middle of April onwards. At the time of emergence the weevils are still immature, and pairing and egg-laying do not take place until about the middle of May. The eggs are deposited singly in green flower heads of clover before the individual flower heads have opened. The duration of the egg stage varies from 4 to 6 days, and the larvae feed on the developing ovules. According to Jenkins (27) there are three larval instars, the total larval period averaging 18 days. The pupal period averages about 6 days and there are two broods in the year. The duration of all the stages is influenced by conditions of temperature and humidity; the larvae are especially sensitive to the latter, being killed rapidly by desiccation of the flower heads during the hay-making process, and this fact has been utilised as a method of control.

Apion trifolii Linn. (*aestivum* Germ.). Clover Weevil

DESCRIPTION. *Adult.* Very similar to the foregoing but smaller, it may be differentiated by the antennae which are all black and the four posterior trochanters which are black instead of reddish as in *A. apricans.* Length 1·5 mm.

The life history and habits of this species are similar to those of the foregoing *A. apricans.*

DISTRIBUTION. Widely distributed in the United Kingdom. In 1925 both species were very plentiful in the eastern and southern counties. In east Devon in 1927 severe injury was caused to young clover plants which were checked by drought.

INJURY TO HOST PLANT. The most important injury is caused by the larvae feeding in the flower heads of clover, on the unripe seeds or in the stem; some damage is also done by the adults feeding on the foliage.

CONTROL. Jenkins (27) finds that the destruction of larvae by desiccation during the haymaking process offers an opportunity of obtaining maximum larval mortality by taking the hay cut at the optimum time. Observations show that when the clover is in the 25 per cent. full flower stage, the larval weevil population is high and a hay cut taken at this stage gives a high percentage of mortality by desiccation.

The recommendations given by Jenkins to farmers growing clover seed are as follows:

(1) Take all the hay cuts, whether of stands destined for seed production or not, when the clover is in 25 per cent. full flower, in preference to taking them at a later date in order to obtain an increase in yield which is practically negligible.

(2) Run the mower over all stands not destined for hay, if poor grazing has resulted in the production of a large number of flower heads. This should be done when the majority of the heads are in the 25 per cent. full flower stage.

(3) In the event of a favourable season causing considerable head production in the aftermath, use the mower in a similar manner.

NATURAL ENEMIES. The Chalcid *Pseudotorymus apionis* Mayr. is recorded from *A. apricans* in Austria, the Braconid *Bracon colpophorus* from *Apion* spp. in France, while the following are recorded from *A. hookeri* in Germany, the Braconids *Bracon satanas* Wesm. and *Aphidius chrysanthemi* Wesm. and the Chalcid *Encyrtus morio* Dlm. A fungus of the genus *Beauveria* sometimes attacks the aerial larvae of *A. apricans*.

REFERENCES

COLEOPTERA

(1) BAER, W. (1920). Die Tachinen als Schmarotzer der schädlichen Insekten. Ihre Lebensweise, wirtschaftliche Bedeutung und systematische Kennzeichnung. *Zeitschr. angew. Entom.* VI, No. 2.

(2) BLUNCK, H. and GORNITZ, K. (1923). Lebensgeschichte und Bekämpfung der Rübenaaskäfer. *Arb. biol. Reichsanst. Land- u. Forstw.* XII, No. 1.

(3) BLUNCK, H. and MERKENSCHLAGER, F. (1925). Zur Oekologie der Drahtwurmherde. *Nachrichtenbl. deutschen Pflanzenschutzdienst*, V, No. 12.

(4) BÖRNER, C. and BLUNCK, H. (1920). Beitrag zur Kenntnis der Kohl- und Rapserdflöhe. *Mitt. biol. Reichsanst. Land- u. Forstw.* Heft 18, 1919.

(5) BÖRNER, C. and BLUNCK, H. (1919). Zur Kenntnis des Kartoffelerdflohs. *Der Kartoffelbau*, III, No. 16.

(6) BURKHARDT, F. and VON LENGERKEN, H. (1920). Beiträge zur Biologie des Rapsglanzkäfers (*Meligethes aeneus* Fab.). *Zeitschr. angew. Entom.* VI, No. 2.

(7) CAESAR, L. (1915). Imported willow and poplar borer (*C. lapathi* L.). *46th Ann. Rept. Entom. Soc. Ontario.*

(8) CAMPBELL, R. E. (1920). *Bruchus rufimanus*. A beetle injurious to beans in California. *U.S. Dept. Agric. Bull.* No. 807.

(9) CARPENTER, G. H. (1903–4). *Reports on injurious insects in Ireland.* XII.

(10) CARPENTER, G. H. (1905–6). A new cabbage-eating larva (*Psylliodes chrysocephala*). *Journ. Econ. Biol.* I.

(11) CHITTENDEN, F. H. and MARSH, H. O. (1920). The western cabbage flea beetle. *U.S. Dept. Agric. Bull.* No. 902.

(12) CURTIS, J. (1860). Farm Insects.

(13) FALCOZ, L. (1929). Un ennemi des cressonnières, *Phaedon cochleariae* F. *Bull. Soc. linn. Lyon*, VIII, No. 3.

(14) FORD, G. H. (1917). Observations on the larval and pupal stages of *Agriotes obscurus* Linn. *Ann. App. Biol.* III, Nos. 2, 3.

(15) FOWLER, W. W. (1887–1913). *The Coleoptera of the British Islands.* 5 vols. London.

(16) FREDERICHS, K. (1920). Untersuchungen über Rapsglanzkäfer in Mecklenburg. *Zeitschr. angew. Entom.* VII.

(17 FRYER, J. C. F. (1917). Insect pests of the basket willow. *Journ. Board Agric.* XXIV, No. 8.

(18) GLENDENNING, R. (1917). The cabbage flea beetle and its control in British Columbia. *Canad. Dept. Agric. Pamph.* N.S., No. 80.

(19) HEIKERTINGER, F. (1915). *Psylliodes affinis.* Abstr. in *Zeitschr. angew. Entom.* II.

(20) HEIKERTINGER, F. (1913). *Psylliodes attenuata* Koch der Hopfen- oder Hanf-Erdfloh. *Verh. d. K.K. Zool.-bot. Ges. Wien*, LXXII, Nos. 3, 4.

(21) HEYMONS, R. (1921). Mitteilungen über den Rapsrüssler *Ceuthorrhynchus assimilis* Payk. und seinen Parasiten *Trichomalus fasciatus* Thoms. *Zeitschr. angew. Entom.* VIII, No. 1.

(22) HODSON, W. E. H. (1929). The bionomics of *Lema melanopa* L. (Criocerinae) in Great Britain. *Bull. Entom. Res.* XX.

(23) HUTCHINSON, H. P. and KEARNS, H. G. H. (1930). Insect pests of willows. *Nature*, CXXV, No. 3145, February 8.

(24) ISAAC, P. V. (1923). The turnip gall weevil. *Ann. App. Biol.* X, No. 2.

(25) JACKSON, D. J. (1920, 1922). Bionomics of the weevils of the genus *Sitona* injurious to the Leguminous crops in Britain, Parts I, II. *Ann. App. Biol.* VII, IX.

(26) JENKINS, J. R. W. (1926). Insect pests of red clover in mid and west Wales. *Welsh Journ. Agric.* No. 2.

(27) JENKINS, J. R. W. (1929). Observations on the control of weevils of the genus *Apion* attacking red clover. *Welsh Journ. Agric.* No. 5.

(28) JENKINS, J. R. W. (1928). Seed treatment as a means of preventing turnip flea beetle attack. *Welsh Journ. Agric.* No. 4.

(29) *Journ. Board Agric.* (1913). *Helophorus rugosus*, XX, No. 1, April, p. 41.

(30) KAUFFMANN, O. (1923). *Arb. biol. Reichsanst. Land- u. Forstw.* XII, No. 3.

(31) KEMNER, N. A. (1916). Gulhariga skirmarbaggen (*B. opaca* L.). *Centralanstalten för Jordbruksförsok. Flygblad* No. 62. *Entomologiska Avdelningen*, No. 15.

(32) KEMNER, N. A. (1917). Artviveln (*Sitona lineatus* L.). *Centralanstalten för Jordbruksförsok. Flygblad* No. 63. *Entomologiska Avdelningen*, No. 16.

(33) LARSON, A. O. (1924). The effect of weevily seed beans upon the bean crop and upon the dissemination of the weevils. *Journ. Econ. Entom.* XVII, No. 5.

(34) LARSON, A. O. (1924). Fumigation of the bean weevils. *Journ. Agric. Res.* XXVIII, No. 4.

(35) LEES, A. H. (1916). *Ann. Rept. Long Ashton Research Station*, p. 37.

(36) LUDWIGS, K. and SCHMIDT, M. (1925). Korbweidenschädlinge. *Biol. Reichsanst. Land- u. Forstw. Flugbl.* No. 81.

(37) MATHESON, R. (1915). Experiments in the control of the poplar and willow borer. *Journ. Econ. Entom.* VIII, No. 6.

COLEOPTERA 147

(38) MEGALOV, B. (1927). *Lema melanopa* L., a pest of oats, barley and other graminaceous plants. *Saratovsk Oblastn S. Kn. Opitn. Stanz. Ent. Otd.* Saratov, 1927.

(39) Ministry of Agriculture and Fisheries, London. Leaflets 10, 25, 143.

(40) MILES, H. W. (1924). A survey of the insect pests of Cruciferous crops. *Kirton Agric. Instit. Bull.* 1. Kirton, Lincs.

(41) MILES, H. W. (1923). A survey of the insect pests of Cruciferous crops. *Journ. Bath and West and Southern Counties Soc.* 5th Series, XVIII.

(42) MILES, H. W. and PETHERBRIDGE, F. R. (1927). Investigations on the control of wireworms. *Ann. App. Biol.* XIV, No. 3.

(43) MUATA, J. and IKEDA, T. (1926). *Lema melanopa* L. in Nagano Prefecture. *Journ. Plant Prot.* XIII, No. 1, Tokyo.

(44) NEWTON, H. C. F. (1928). The biology of flea beetles (*Phyllotreta*) attacking cultivated Cruciferae. *Journ. S.E. Agric. Coll. Wye, Kent*, No. 25.

(45) NEWTON, H. C. F. (1929). Observations on the biology of some flea beetles of economic importance. *Journ. S.E. Agric. Coll. Wye, Kent*, No. 26.

(46) PARKER, W. B. (1910). *Psylliodes attenuata. U.S. Dept. Agric. Washington, Bull.* No. 82.

(47) PETHERBRIDGE, F. R. (1928). The turnip mud beetles. *Ann. App. Biol.* XV, No. 4.

(48) RÉGNIER, R. (1928). Les taupins nuisibles en grande culture. Contribution à l'étude de l'*Agriotes obscurus* L. *Rev. Path. vég. Ent. Agric.* XV, No. 2, Paris.

(49) ROEBUCK, A. (1916). An attack of mustard beetles on watercress. *Journ. Board Agric.* XXIII, No. 3.

(50) ROEBUCK, A. (1928). Studies in the genus *Phaedon. Lincs. Nat. Union Trans.* 1927. Lincoln.

(51) ROSTRUP, S. (1920). Jordloppeangrebet i 1918. Jordloppernes levevis og forsog med deres bekaempelse. (Flea beetle attack in 1918. The habits and control of flea beetles.) 142 Beretning Statens Forsogvirks i Plantekultur. *Tidsskrift for Planteavl*, XXVII. (With a summary in English.)

(52) RYMER ROBERTS, A. W. (1920–28). On the life history of wireworms of the genus *Agriotes* Esch. With some notes on that of *Athous haemorrhoidalis* F. Parts I–IV. *Ann. Appl. Biol.* VI, VIII, IX, XV.

(53) SCHEIDTER, F. (1926). Forstentomologische Beiträge. *Zeitschr. f. Pflanzenkrank.* XXXVI, Nos. 5–6.

(54) SKAIFE, S. H. (1918). Pea and bean weevils. *Union of South Africa Dept. Agric. Pretoria Bull.* No. 12.

(55) SMITH, Kenneth M. (1925). Further experiments in the control of certain maggots attacking the roots of vegetables. *Ann. App. Biol.* XII.

(56) TAYLOR, T. H. (1918). Observations on the habits of the turnip flea beetle. *Entomologist*, LI, London.

(57) TAYLOR, T. H. and THOMPSON, H. W. (1928). A garden chafer attack. *Ann. App. Biol.* XV, No. 2.

(58) THEOBALD, F. V. (1903). *Report Econ. Zool. Brit. Museum Nat. Hist.* p. 8.

(59) THEOBALD, F. V. (1924). *Report S.E. Agric. Coll. Wye, Kent.*

(60) THEOBALD, F. V. (1913). *Report on economic zoology for the year* 1913. S.E. Agric. Coll. Wye, Kent.

10-2

148 COLEOPTERA

(61) THEOBALD, F. V. (1925). *Cultivation, diseases and insect pests of the hop crop.* Min. Agric. Miscell. Publ. No. 42.

(62) THEOBALD, F. V. (1911). *Report on economic zoology.* S.E. Agric. Coll Wye, Kent, p. 85.

(63) TÖLG, F. (1913). *Psylliodes attenuata* Koch der Hopfen- oder Hanf-Erdfloh. Part 1. *Verh. d. K.K. Zool.-bot. Ges. Wien,* LXXII, Nos. 3, 4.

(64) TÖLG, F. (1915). *Psylliodes affinis* Payk. Abstract in *Zeitschr. angew. Entom.* XI, pp. 1–9.

(65) TOLTZ and HEIKERTINGER F. (1916). *Psylliodes affinis* Payk. der Kartoffel-erdfloh. *Zeitschr. f. Pflanzenkrank.* XXVI, Nos. 6, 7.

(66) VASSILIEV, E. M. (1913). *The chief remedies against the larvae and beetles of* Lema melanopa *L. a pest of summer sown grain.* Studies from the Experimental Entomological Station of the all-Russian Society of Sugar Refiners for 1912. Kiev. (In Russian.)

(67) Wireworms (1914). *Journ. Dept. Agric. and Techn. Instr. Ireland,* XIII, No. 4.

CHAPTER X

HYMENOPTERA

Order 21. **Hymenoptera.** Sawflies, Ants, Bees, Wasps

Two pairs of membranous wings, venation reduced or absent; hind wings smaller than fore wings; mouth parts biting or sucking; ovipositor always present, modified for stinging, piercing or sawing; metamorphosis complete.

Three members of this order are of some importance as agricultural pests and are dealt with briefly. These three insects belong to the sub-order Symphyta (super-family Tenthredinoidea), the sawflies, and are distinguished from the remainder of the Hymenoptera by the absence of the constriction between thorax and abdomen and by the adaptation of the ovipositor for boring or sawing.

Family *Tenthredinidae*

Athalia colibri F. (*spinarum* F.). The Turnip Sawfly

DESCRIPTION. *Adult.* General body colour bright orange; antennae 9-jointed, black above, yellowish beneath except at the base and apex, the joints decreasing in length but increasing in diameter towards the apex. Head and eyes black, mouth white with a whitish pubescence, three ocelli triangularly placed. Thorax with three orange marks, scutellum yellow. Abdomen in female broad, flattened with partly concealed ovipositor, sheath black at apex and hairy; in male somewhat cylindrical. Legs stout and short; tibiae hairy; tarsi whitish; apex of tibiae and joints of tarsi black. The male has the two basal joints of the antennae pale yellow, in the female the antennae are often brownish underneath. (Fig. 37.) Length 6–8 mm.

Egg. Very small, oval, whitish and semi-transparent.
Larva. On hatching, the larva is white with two black dots on the head, later the body becomes darker and the head black. When full grown the head is narrower than the second segment, shining black with a few short hairs. Each of the body segments is divided into several folds, smooth and shining without any hairs. The upper part of the spiracles is black, on each side is a longitudinal slate-coloured spot, then a row of black mostly double, oblong

spots. Legs, slate-coloured, abdominal legs eight pairs, black splashed with grey, almost hidden by the overhanging folds of the body. (Cameron[2].)

Pupa. Greyish white in an oval cocoon formed of grains of earth.

LIFE HISTORY. The adults appear in May and the female deposits her eggs singly along the leaf margin on the under side, embedded in the epidermis. From two to three hundred may be laid by one female, arranged along the leaf edge at irregular intervals. The eggs

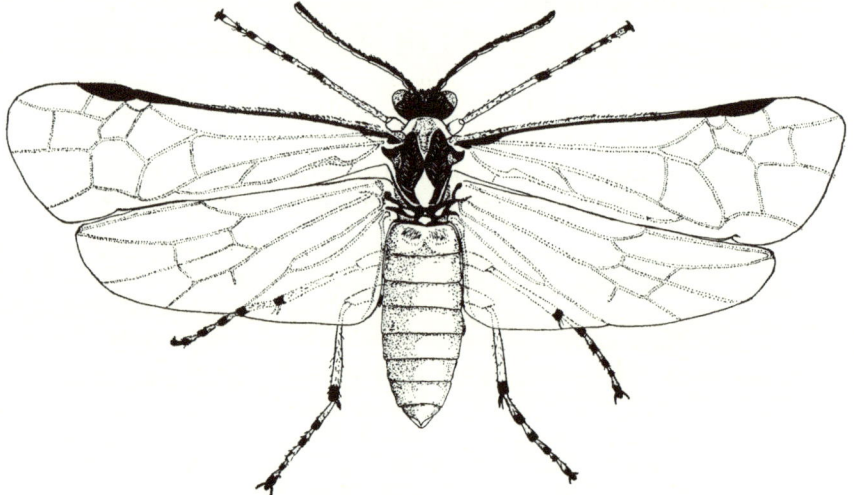

Fig. 37 *Athalia colibri* F. The Turnip Sawfly. × 10.

hatch in 5–12 days according to the temperature, and the young larvae feed at first gregariously on the leaf epidermis, forming a brown patch; later they skeletonise the leaf. There are three larval moults at intervals of 5–7 days, the total length of the larval stage being about 19 days. When full grown the larvae descend into the soil where they pupate, forming a silken cocoon covered with soil particles. The pupal stage of the early summer brood lasts about 21 days and there are usually three broods in the year; the first in early summer, the second at the end of July and beginning of August, while the third brood feeds as late as October; pupation then takes place in the soil, where the winter is passed.

Host plants. The host plants consist of cultivated Crucifers, especially turnips, swedes, and white mustard (*Sinapis alba* Linn.), also various Cruciferous weeds such as charlock (*Sinapis arvensis*), winter cress (*Barbarea* spp.), and hedge mustard (*Sisymbrium officinale*). This insect is not usually a serious pest, but epidemics occasionally develop when the larvae appear in such numbers as to destroy the crop. These epidemics have not appeared of late years, a fact possibly due to crop rotation.

Control. In the case of a bad attack, poison sprays to destroy the larvae might be used, such as lead arsenate or soap-pyrethrum, while dusting the plants with kainit is said to be efficacious. Deep ploughing after the crop is harvested, to destroy the pupae, is also recommended.

Distribution. General all over the south of England, but more rare in the north and in Scotland.

Natural enemies. The larvae are parasitised by the thread worm *Mermis albicans*. The Tachinid flies, *Meigenia bisignata* Meig., *Lydella nigripes* Fall. and *Tachina erucarum* Rnd., and the following Hymenoptera are recorded from Europe:

Ichneumonidae
  Bassus athaliaeperda Curt.
  Tryphon succinctus Gr.
  T. brachyacanthus Gr.
  T. marginellus Gr.
  Perilissus lutescens Holmgr.
  Mesoleius armillatorius Gr.
  M. ciliatus Holmgr.
  M. filicornis Holmgr.

Ichneumonidae
  Mesochorus areolarius Ratz.
  Spanotecnus lutescens Thoms.
  Exenterus succinctus Gr.

Chalcididae
  Perilampus splendidus Dolm.
  P. ruficornis F.
  P. violaceus

Cephus pygmaeus Linn. Corn Sawfly; Wheat Stem Sawfly

Description. *Adult*. (Fig. 38.) Shiny black. Head rather large; eyes prominent, three ocelli on the crown; antennae rather long and slender with 21 joints, slightly clubbed, inserted in front of face. In the male the mouth and mandibles are bright yellow, with a spot of yellow on the clypeus, interior margin of eyes yellow. Thorax oval not broader than head. Abdomen sessile, long, slender, slightly compressed; at the base a yellow membranous spot, there are yellow spots on each side of the first and second segments and a dot on the back of the latter, third and fifth segments have broad yellow margins, the sixth has a narrow one forming spots on the sides and back, apex yellow. Wings transparent, iridescent; nervures brown and slender. Legs bright yellow, including coxae and trochanters, but they as well as the femurs have black stripes on the outside.

Female, rather darker, with palpi and sides of jaws only yellow; abdomen shorter and stouter, yellow spots on the two basal joints reduced or absent; ovipositor black; legs ochreous; but coxae, trochanters and femora black, hinder tibiae brown on the inside surface. (Curtis(4).) Length 8–10 mm.

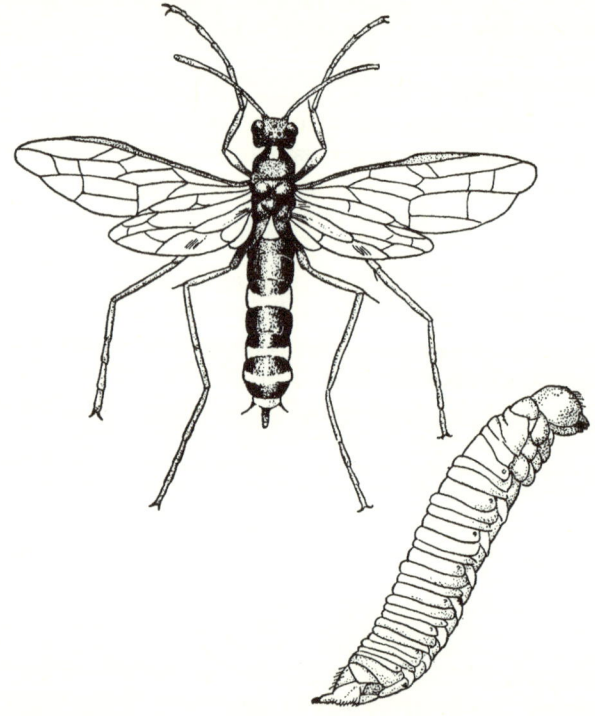

Fig. 38. *Cephus pygmaeus* Linn. The Wheat Stem Sawfly and larva. × 4.
(After Criddle.)

Egg. Shiny, almost transparent, somewhat crescent-shaped when first deposited, later becoming reniform. Length 1·1 mm.

Larva. Almost cylindrical in shape, dull yellowish white; head, rounded, horny and brown in colour; mandibles large and toothed; antennae 4-jointed and tapering; thoracic segments swollen, larger than the head, the legs are absent but are represented by short tubercles, there are ten pairs of spiracles, two on the thorax and eight on the abdomen, the prothoracic spiracles are larger than the others and elongate; the last abdominal segment terminates in a tubular appendage, capable of extrusion, this is probably used to assist progression through the stem of the host plant. There is a stout spine on each side of the ventral lobe at the caudal end; body devoid of hairs except the head and caudal end. (Ries (8).) Length 10–14 mm. (Fig. 38.)

LIFE HISTORY. The adults appear in June and the females oviposit in the stems of wheat and barley, tall thick-stemmed plants are preferred for oviposition as are also those in the ear-forming stage. The method of oviposition is as follows, the female rests head downwards on the stem, draws the abdomen well under the body and thrusts the ovipositor into the leaf sheath at or below a joint. The egg is usually inserted in the tissue but may be laid on the outside, one egg being laid on each stem. The eggs hatch in 7–10 days and the young larva lives inside the straw, feeding on the pith, usually low down in the stem. When full grown the larva descends and cuts round the stem, shortly above the ground level; the stem is either severed cleanly or may be broken off by the wind. The larva then blocks up the end with debris and excrement, and passes the winter in the short length of straw below the cut. (Fig. 39.) It pupates early in the following summer and emerges as an adult in June. There is only one brood in the year. Attacked stems may first be recognised by the white ears which stand erect as compared with the normal growing corn, and later by the severed stems. In harvesting, the hibernating larvae are left behind in the stubble, so that control measures should aim at the destruction of this, which would be best accomplished by burning. The varieties of wheat seem to vary little in their susceptibility to attack. Time and density of sowing

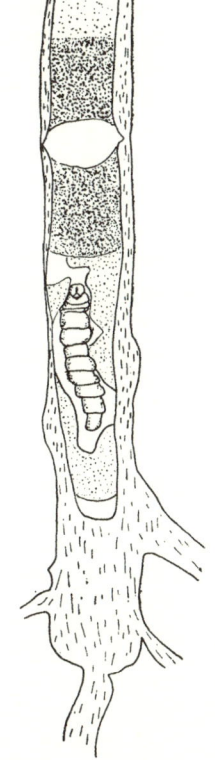

Fig. 39. Larva of *Cephus pygmaeus* Linn. *in situ* in stem of wheat. Note the plug made by the larva above and below the incision where the stem finally breaks off. About twice natural size. (After Ries.)

affect the amount of subsequent infestation by their influence upon the condition of the crop at the time when the sawflies are in flight.

CULTIVATED HOST PLANTS. Chiefly wheat, barley and occasionally rye; oats apparently are not attacked.

WILD HOST PLANTS. Reported as infesting meadow grasses in Germany; Ries in America records it from a species of *Bromus*. The adult insects appear to show a preference, when feeding, for flowers with a yellow or blue colouring.

NATURAL ENEMIES. There are records of two Hymenopterous parasites from *Cephus pygmaeus* in Europe; these are the Ichneumonid *Collyria calcitrator* Gravenh. and the Chalcid *Arthrolysis scabricula* Nees.

In addition the following have recently been bred from *C. pygmaeus* in England by Salt, to whom the writer is indebted for the records: the Ichneumonids, *Pimpla detrita* Holmgr., *Hemiteles hemipterus* Fabr. and *Collyria calcitrator* Grav., the Braconid, *Microbracon terebella* Wesm. and the Chalcid *Pleurotropis benefica* Gahan.

Ries (8) records the following from *Cephus pygmaeus* in America: the Chalcids, *Eupelmus allynii* French, *Eupelminus saltrator* Lind., *Pleurotropis benefica* Gahan, *Eurytoma* spp., the Braconid *Heterospilus cephi* Rohw. and the Ichneumonid *Hoplocryptus* sp.

Pontania proxima Lep. (*gallicola*) Steph. Willow Gall Sawfly

DESCRIPTION. *Adult.* Black, shining. Clypeus, labrum, tegulae, the pronotum close to the tegulae and legs, whitish yellow; the greater part of the coxae and a more or less longish line on the femora black; base of posterior tibiae and the tarsi fuscous; antennae a little longer than the abdomen, apical joints often brownish beneath, sutures on vertex short; antennal depressions large, deep. Wings hyaline; costa pale; stigma a little longer than broad, almost quadrate.

Female. Antennae longer, thicker and more pilose; flagellum brownish; stigma almost wholly fuscous; wings not so clearly hyaline; femora with more black. (Cameron (2).)

The male is rarely seen and the species is parthenogenetic. Length 3–4 mm.

Egg. Opaque, oval, bluish white in colour. Length 0·5 mm.

Larva. When newly hatched, the larva is white and almost transparent with a brownish black shining head. The mature larva is greenish white in colour, the green tint being probably due to ingested food. The head is dark and shining with a whitish area on the face; mouth pale brown; mandibles darker; legs white with brown claws. Length 6–8 mm.

LIFE HISTORY. The mature larva passes the winter in a cocoon and pupates in spring; the adults appear in May and oviposit in the leaf buds of willows on the outer or under side of the leaf, and the characteristic bean-shaped gall grows with the larva in the leaf, one larva is found in each gall. The galls are often placed at the edges of the leaves, and are green at first, later turning red with a paler tint below; any number of galls from one to twelve may be found on one leaf. When full fed the larva eats its way out of the gall and falls to the ground where it pupates, forming a white pupa in a cocoon in the soil. Occasionally the cocoons are to be found in crevices in the bark of the tree. There are two broods in the year, the second appearing in July and August and ovipositing in the mature leaves. This brood is often heavily parasitised, but there appear to be no records of the identity of these parasites. Murphy, I. M. (*Entom. Mon. Mag.* LXV. Dec. 1929) gives an account of the methods of egg-laying of this sawfly.

REFERENCES

HYMENOPTERA

(1) AINSLIE, C. N. (1929). The western grass-stem sawfly, a pest of small grains. *U.S. Dept. Agric. Tech. Bull.* No. 157.

(2) CAMERON, P. (1889). *A Monograph of the British Phytophagous Hymenoptera.* Ray Society, I.

(3) CRIDDLE, N. (1915). The western wheat stem sawfly. *Dom. of Canada Dept. Agric. Entom. Div. Bull.* No. 11.

(4) CURTIS, J. (1860). *Farm Insects.*

(5) GAHAN, A. B. (1920). The black grain-stem sawfly of Europe in the United States. *U.S. Dept. Agric. Bull.* No. 834.

(6) MILES, H. W. (1923–4). *Journ. Bath and West and Southern Counties Soc.* 5th Series, XVIII.

(7) NOEL, P. (1915). Les ennemis de l'avoine. *Lab. Ent. Agr. Seine-Inf. Bull.* Trimest. I, Nos. 4–7.

(8) RIES, D. T. (1926). A biological study of *Cephus pygmaeus* Linn., the wheat-stem sawfly. *Journ. Agric. Res.* XXXII.

(9) THEOBALD, F. V. (1913). Report on economic zoology. *Journ. S.E. Agric. Coll. Wye, Kent.*

(10) TORKA, V. (1928). Ein Schädling des weissen Senfs (*Sinapis alba* L.) *Anz. Schadlingsk.* IV, No. 4.

CHAPTER XI

DIPTERA

Order 22.. **Diptera**. True Flies

One pair of membranous wings attached to mesothorax, the second pair represented by halteres or 'balancers' borne by the meta-thorax. Mouth parts suctorial, sometimes adapted for piercing. Larvae legless. Metamorphosis complete.

This is a very large order of insects and is one which closely affects man in many spheres of his activity.

Sub-order A. *Orthorrapha*

Series I. *Nematocera*

Family *Tipulidae*. 'Daddy-long-legs'; Crane Flies

Large flies, antennae long, many-jointed, legs long and fragile. Head prominent, rounded without ocelli, there is a V-shaped suture present on the dorsal surface of the mesothorax. Ovipositor horny and valvular.

Tipula paludosa Meig. Marsh or Allied Crane Fly

DESCRIPTION. *Adult*. (Fig. 40.) Male. General colour grey with a reddish brown tinge. Head ashy, proboscis tawny, palpi brown; antennae 13-jointed, dark, first two segments testaceous. Thorax reddish brown, somewhat raised with four indistinct stripes, meta-thorax and pectus whitish. Halteres whitish and rather long. Abdomen ferruginous to reddish yellow, dorsal stripe faint, up-turned at the extremity. Wings rather narrow with a reddish tinge, brown along the costa, an *indistinct* pale streak under costa. Legs brown, rather stout for a Tipulid; femora and tibiae blackish at tips; tarsi dark.

Female. Wings *shorter* than abdomen which is long and pointed, ovipositor pale brown, no wing streak. Length 17–25 mm.

Egg. Black, unsculptured, with a purplish metallic lustre, spindle-shaped, somewhat asymmetrical one end being slightly conical, the other rounder. Oldham (8) considers the micropyle to be represented by a small circular patch near the pointed end. Size 1·1 × 0·4 mm.

Larva. Much of the information concerning this larva is derived from the work of Rennie (9). The young larva is pale sandy red in colour, and the 13th segment bears a pair of tufts of relatively strong curved bristles, these tufts constitute a point of difference between the early and late stage larvae. The full-grown larva is brownish grey in colour with numbers of irregularly placed black dots. In shape it is cylindrical, slightly narrowed anteriorly and expanded posteriorly. The tough, usually tense, skin has numerous transverse wrinkles, with the segments marked off by slight furrows, each segment bearing on its ventral surface four minute bristles. The head bears a pair of short jointed antennae and a strong chitinous mouth armature. The anus is surrounded by large fleshy lobes and

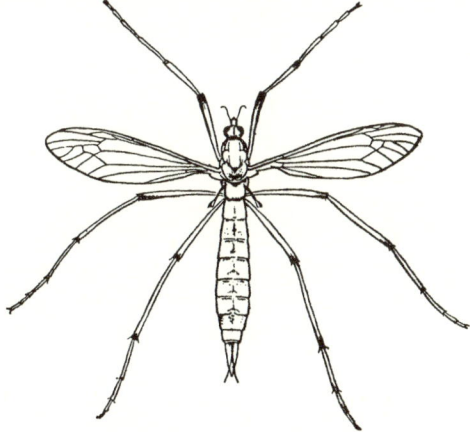

Fig. 40. *Tipula paludosa* Meig. Marsh or Allied Crane Fly. × 1½.

there is a pair of large retractile laterally placed papillae. The terminal segment is somewhat truncated and bears the posterior spiracles which are circular and brown in colour; bordering this stigmatic area are six conical papillae of definite form and arrangement, these papillae are of importance in determining the species in Dipterous larvae generally. In this larva there is a ventral pair with black tips and a clear central area, below each of these is a pair of pigmented spots. The remaining four papillae project dorsally in two pairs. (Fig. 41.) On account of the toughness of the skin, these larvae are universally known as 'leatherjackets'. Length 40 mm.

Pupa. Brown in colour, rather stout, cephalic horns straight, slightly recurved at tip, wing sheaths reaching to about second abdominal segment, leg sheaths to 4th–5th. There is a ventral row of spines situate basally on the posterior abdominal segments;

cauda with dorsal armature of four lobes, while on the eighth segment there are ten spines, two lateral, four ventral and four dorsal. The lateral abdominal spiracles are reduced or absent. Length 20–25 mm.

LIFE HISTORY. The winter is spent in the ground in the larval state, the larvae occasionally feeding during mild weather. The adults appear in June and mating and oviposition take place soon after emergence, the eggs being deposited in the ground just below

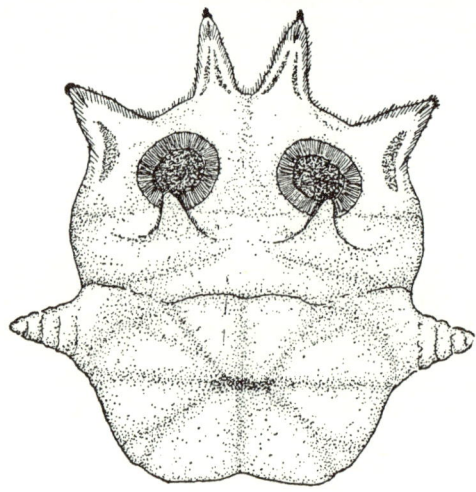

Fig. 41. Larva of *Tipula paludosa* Meig., posterior aspect of segment 12 showing the spiracular disc and the anal area. × 14. (After Oldham.)

the soil surface. The act of oviposition is performed in a characteristic manner; the female, holding her body in a vertical position, penetrates the ground with her ovipositor by means of a series of rotating movements; each female is capable of laying about 300 eggs. The eggs hatch in 14 days and the larvae live for nine months, roughly from September to July, feeding voraciously upon various crops; pupation takes place in the soil, the pupal period being 10–14 days. According to Rennie there is only one brood per annum in north-east Scotland and the same is probably true in England although in late autumn the development of a partial second brood sometimes occurs. The writer has observed the appearance of large swarms of this and the succeeding species in mild weather during October.

Tipula oleracea Linn. The Common Crane Fly

DESCRIPTION. *Adult.* Male. General colour silvery or dusty grey. Head grey, tawny behind; eyes black; proboscis brownish; antennae 13-jointed, dark, at least the first three segments testaceous; palpi brown. Thorax brownish with four pale indistinct brown stripes; metathorax and pectus greyish white; halteres brown, blackish at ends. Abdomen slaty grey, testaceous near tip which is upturned. Wings greybrown, clear, wing streak whitish and rather *distinct*. Legs brownish; femora and tibiae darker at tips; tarsi black.

Female. Similar, wings larger; *longer* than abdomen which is tawny at the tip. Length 15–23 mm.

Egg. Black, shining, surface reticulate, elongate-oval in shape, somewhat asymmetrical. Size 1 × 0·3 mm.

Larva. Colour a uniform earthy grey; rather conical in shape towards head and rounded posteriorly; head black; antennae prominent; mandibles conspicuous. Length 40 mm. when fully extended. (Fig. 42.)

Fig. 42. Larva and Pupa of *Tipula* sp. × 4.

Pupa. Pale in colour at the time of transformation, darkening to blackish brown, darker on the head and back of thorax. There is a yellow marginal streak extending along the thoracic and abdominal segments and also a double line of light grey crossing the segments dorso-ventrally. The abdominal spines are situated on the posterior margin of each segment, those on the dorsal surface being smaller, the caudal end is spiny and pointed. (Fig. 42.) Length 20–25 mm.

LIFE HISTORY. Very similar to the foregoing *T. paludosa*, the adults appear earlier and may be seen on the wing from May to August. There is probably only one brood in the year as a general rule with an occasional partial second one.

POINTS OF DIFFERENCE BETWEEN *T. PALUDOSA* AND *T. OLERACEA*

The following distinguishing characters are given to aid identification of these two closely allied and very similar flies.

T. paludosa. Colour reddish; pale streak on wing *indistinct*; first two antennal segments only testaceous; abdomen of female reddish yellow and more elongate; wings in the female narrower and *shorter* than the abdomen.

T. oleracea. Colour greyish; pale streak on wings *distinct*; the first three antennal segments, at least, testaceous; abdomen of female grey; wings in the female *longer* than the abdomen; slightly smaller insect.

To these points Britten(3) makes the following additions: *T. oleracea*—antennae 12-jointed, eyes almost touching beneath head; *T. paludosa*—antennae 13-jointed, eyes widely separated beneath head. The same writer, quoted by Oldham(8), mentions the existence of a third species, *Tipula czizeki* Jong., which closely resembles *T. oleracea* but can be differentiated from it by the wholly dark antennae and the eyes widely separated beneath head. Morrison(7), also, gives the microscopical differences existing in the male genitalia of these two flies.

Pachyrrhina maculata (*maculosa*) Meig. Spotted Crane Fly

DESCRIPTION. *Adult.* (Fig. 43.) Male. General body colours yellow and black. Head yellow with a triangular black spot on occiput, two small black spots on vertex; proboscis black dorsally, yellow laterally; antennae, eyes and palpi black. Thorax yellow with three black stripes, the centre one short, outer pair converging at their base; scutellum yellow-brown; halteres testaceous. Abdomen yellowish brown with an interrupted dorsal black stripe, terminal segments upturned, some black markings laterally; there is a short yellowish pubescence. Wings greyish, veins brown, stigma yellowish. Legs, femora and tibiae brown darkening towards joints, front femora darker, tarsi darker.

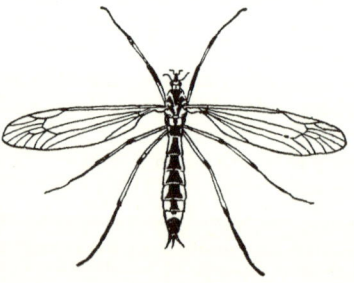

Fig. 43. *Pachyrrhina maculata* Meig. The Spotted Crane Fly. × 1½.

Female. Similar, abdomen longer and more cylindrical, the dorsal abdominal stripe rather more interrupted. Length 12–20 mm.

Egg. Oval, black and shining, somewhat spoon-shaped.

Larva. Smaller than the two foregoing species, greyish brown in colour, skin wrinkled but less tough than in *Tipula.* Shape cylindrical, somewhat tapering to either end, the alimentary canal is visible through the integument. There is a transverse row of stiff bristles on each segment with some black hairs laterally. These larvae can be distinguished from the preceding species by the arrangement of the tubercles on the caudal corona. In *P. maculata* there are two papillae with two shorter ones between them, two blunt tubercles below and two fleshy protuberances which are capable of dilatation. There are also the two central posterior spiracles, and between these and the ventral papillae are three small dark spots.

Pupa. Brown to light brown in colour, narrow with two straight processes or breathing horns on the head. Abdominal segments each with a transverse row of small spines above and five larger ones below; there is also a lateral row of spines; the penultimate segment bears six long spines and two shorter ones. There is a large conical process at the end of the abdomen, and a shorter one beneath it.

LIFE HISTORY. The bionomics of this species are similar to those of *T. oleracea.* The larvae pupate in the spring and the adults fly from May to August. *P. maculata* is commonly met with in gardens, and is perhaps more frequently found attacking garden produce than are the *Tipula* species.

CROPS ATTACKED AND INJURY TO HOST PLANT. The larvae of one or more of the foregoing three species of flies are almost always present in farm lands, particularly in pasture and meadow land which are their natural habitat. In addition they attack a multitude of agricultural crops, particularly oats, which in some years suffer severely; turnips, mangolds and potatoes are also very frequently attacked. The worst damage is done in the spring when the plants are in the seedling stage, especially in the case of oats before the establishment of the adventitious roots. The injury is mostly done below ground by the larva gnawing at the roots and underground portions of the stem. Very often, in the case of young Brassicas, the stem is severed at soil level much after the manner of cutworm attack (p. 73).

CONTROL. On arable land the best remedy is undoubtedly a poison bait made up as follows: 20 lb. of bran, 1 lb. Paris green, 1 gallon of water and ½ pint of molasses or treacle; this amount is sufficient for one acre. The mixture should be broadcast so that it is thinly and evenly distributed and the cost should not exceed 6s. per acre. This method has been widely tested and has given satisfactory control, no case of poisoning of birds having been reported. In 1927 the Protection of Animals Act (1911) was amended to allow the application of such poison baits on arable lands, under 'reasonable precautions'. In the writer's experience with this poison bait, the substitution, sometimes recommended, of lead arsenate for Paris green has rendered the bait ineffective, probably because the former is much less soluble than the copper salt. Gasow(5) recommends application to the soil of a solution of ammonium carbonate or a 2–4 per cent. solution of ammonia, which appears to bring the larvae to the surface. As substitutes for the Paris green bait, the following are sometimes recommended as being cheaper and almost as effective: sodium fluoride, 1 part to 25–40 parts of bran; or sodium fluosilicate, 1 part to 50 parts of bran. On grass land where the larvae do not come up to the surface to feed, the poison bait is not applicable, and in this case Gammexane or D.D.T. should be used. On turf, a 5 per cent D.D.T. powder at the rate of ½ oz. per square yard, or alternatively a 5 per cent. emulsion, 1 gallon diluted with water to 400 gallons, are recommended.

If Gammexane is used, 1 oz. per square yard containing 3½ per cent. benzene hexachloride is best (Dawson and Escritt (1946), *Nature, Lond.*, CLVIII, 448.) These methods of control apply equally to all three species of crane fly. Leatherjackets, particularly when newly hatched, are very susceptible to drought and in consequence farmers consider that a hot and dry August and September reduce the numbers of the overwintering larvae, while a wet autumn is favourable to their development, and some confirmation of this is forthcoming from the report of the Ministry of Agriculture on insect pests for 1925–7.

NATURAL ENEMIES. Little appears to be known of the Hymenopterous parasites of Tipulids. Rennie records a Tachinid *Bucentes geniculata* from the larva of *T. paludosa*, and the three Tachinids *Admontia amica* Meig., *A. blanda* Fall., and *Bucentes cristata* F. are recorded from species of *Tipula* in Europe. Birds, including poultry, play a large part in the natural control of leatherjackets, particularly rooks, starlings, plover, various gulls and sometimes pheasants. A Mermithid worm has been described from Scotland as parasitising the larva of *T. paludosa* (Rennie(11)) and a fungus *Entomophthora arrenoctona* is recorded from France as attacking *T. paludosa*.

Family *Tabanidae*. Horse Flies; Clegs; Breeze Flies; Gad Flies; etc.

Mostly large robust insects without large bristles; head convex before, concave behind; antennae short, 3-jointed; eyes large, close together in male, widely separated in female, showing rainbow colours in life, proboscis projecting, adapted for piercing and blood-sucking in the female.

Several species of this group annoy livestock and suck their blood. Only one species, *Haematopota pluvialis* Linn., is described in detail. Others which may be mentioned are *H. crassicornis* Whlbg., *Tabanus sudeticus* Zlr., *T. bovinus* Linn. and *Chrysops relictus* Meig.

Haematopota pluvialis Linn. The Dun Fly or Cleg

DESCRIPTION. *Adult*. Male. Head broader than thorax, triangular, frons bare, yellowish with two large black spots and a small median one, the middle part of frons jet black; cheeks whitish with small black spots, coalescing near base of antennae, pubescence silvery; antennae brown, first joint dilated, second cup-shaped, third joint with basal segment long. Thorax brownish black, dull, with three narrow grey lines, pubescence pale yellowish, long; scutellum dark with some long black and grey bristles. Abdomen dull black with sutures brown or greyish. Wings marbled with hyaline spots on a greyish background, and a profusion of ring-like or scroll-like markings. Legs dull black, alternating with rings of yellow or orange.

Female. Similar, frons broader than eye, thorax greyish black not dull black as in male, with five yellowish grey lines; scutellum with pale pubescence. Length 8–10 mm.

LIFE HISTORY AND HABITS OF TABANIDS. The eggs are deposited in rounded or flattened masses on leaves of aquatic plants or other smooth surfaces near water. Cameron gives an account of the egg-laying habits of *H. pluvialis*. (*Nature*, 126, No. 3181, Oct. 1930.) Tabanid larvae are whitish and occur in the water or soil, they are carnivorous and feed upon the small invertebrate fauna of these localities. They are cylindrical in shape, tapering towards each end, with a small retractile head and with the first seven of the eight abdominal segments each encircled near its anterior margin with a ring of fleshy protuberances, of which there are two transverse dorsal, one lateral on each side, and four rounded ventral ones. (Austen (13).) It is probable that the larger species (*Tabanus*) may require two years to complete its life cycle. The biting habits of the adults vary, some bite the abdomen almost exclusively, others prefer the face and different parts of the body. As a rule the smaller species are the most annoying to stock. Horses suffer more from the attacks of these flies than cattle which, being heavier in the hide, are often indifferent to the bites.

PROTECTION OF STOCK. Hadwen (15) recommends nets and canvas coverings as a great help to horses when they are at work. Fly repellents seem to be of little use. Darkened shelters are useful and are only required during the heat of the day, because the flies are not active till late in the morning and disappear about sunset.

NATURAL ENEMIES OF *TABANIDAE*. The Proctotrypid egg parasite *Phanurus emersoni* Gir. is recorded by Parman (16) in America; Cameron (14) records the same, and also the Chalcid egg parasite *Trichogramma minutus* Riley from Tabanids in Canada.

Family *Cecidomyidae*. Gall Midges

Minute and delicate flies, body and wings clothed with long hairs; antennae long, moniliform, each segment usually with a whorl of hairs; wings hairy with few longitudinal veins, cross veins absent; legs long and slender.

Mayetiola destructor Say. Hessian Fly

DESCRIPTION. *Adult. Female.* Head, eyes, frons and occiput black; palpi yellow; proboscis very small and pink; antennae slightly more than one third length of body, yellowish brown with 17 joints, the two basal joints are nearly twice as thick as the others. Thorax black with two indistinct lines of white hairs along the dorso-central region, a reddish irregular streak runs along the lower side of the thorax to the base of the wing; scutellum black, halteres pale red with black scales. Abdomen pinkish or yellowish brown, with eight segments, all except the first marked on each side of the dorsum with a square black spot, the lines of spots converge on the 7th–8th segments, a similar line of spots runs down the centre of the ventral surface; ovipositor pale red. Legs pinkish; coxae brown; trochanters black. Wings pink at the base, clothed with black hairs.

Male. About one third shorter than the female and more slender, antennae 17-jointed, pedunculate, two thirds length of body. Abdomen almost black with a pink extremity, black markings similar to female but they coalesce owing to the slenderness of the body. The male organs are conspicuous and consist of a pair of claspers of a brown colour between which are situated the pink generative organs. Legs are paler than those of the female, while the wings are longer and less black. (Shortened after Meade (39).)

Egg. Oblong, cylindrical, glossy, of a reddish colour. Length 0·5 mm.

Larva. Colour variable, yellowish white to pale green, sometimes tinted with red, rather translucent. In shape, flattened, tapering slightly to either end, somewhat slug-like. Legless.

Pupa. Semi-oval, tapering to a point, reddish brown in colour. Often known as 'flaxseeds' from a superficial resemblance to such, but in reality the pupa is smaller, narrower, darker and less flat. Length 6 mm.

LIFE HISTORY. The Hessian fly, so called in America because of its supposed importation into that country from Germany by Hessian troops, is a very serious pest of wheat both in America and Southern Europe. The life history as briefly described here is applicable only to Great Britain where, on account of the late sowing of the autumn wheat, the more serious autumn attack of the fly does not develop. Partly for this reason and partly because climatic conditions are not suited to it the Hessian fly is seldom a very serious pest. The adult flies appear in May from the overwintering pupae, oviposition commences very shortly after hatching and the eggs are laid on the leaves, usually along the creases on the upper surface, and according to some authorities on the stem itself. On hatching from the egg the larva migrates towards the leaf sheath; as the eggs are usually laid with the anterior end pointing to the leaf tip it follows that the larvae must turn round to reach the leaf sheath. If, as sometimes happens, the eggs are laid with the anterior end facing the leaf base, the larvae crawl up the leaf to the tip and turn round. This migration of the larvae appears to be productive of a high mortality. The eggs hatch in 4–8 days and the young larvae make their way under the sheathing leaf where they feed upon the sap of the stem, usually above the first or second joint. The larvae are mature in about 20–30 days and pupation takes place *in situ* between the leaf sheath and the stalk. There is only one main brood of the Hessian fly in England but probably a small percentage of the pupae hatch in September and these flies deposit their eggs on wild grasses. The majority remain in the stubble as pupae or are harvested and dislodged from the straw by threshing. The winter is thus passed in the pupal condition in the old stubble or in the screenings from the threshing machine.

CULTIVATED HOST PLANTS. Primarily a wheat insect, but it also occasionally attacks barley and rye, very rarely oats.

WILD HOST PLANTS. Probably only couch grass (*Agropyron repens*); there seems to be some doubt as to whether the Hessian fly does breed on the other wild grasses.

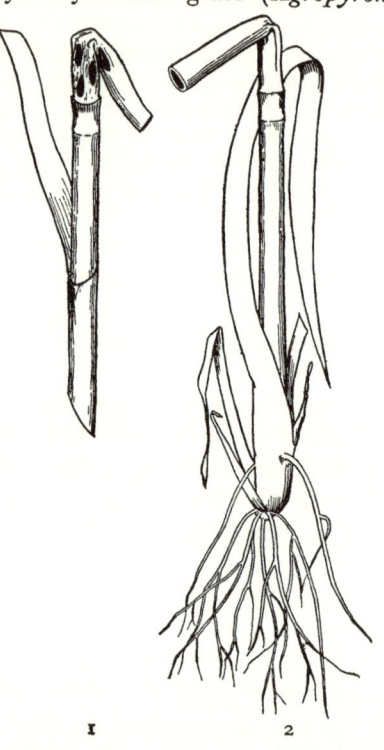

INJURY TO HOST PLANT AND SYMPTOMS OF ATTACK. The stems are weakened and dwarfed, the crop beginning to fail in June and July, the most marked symptom is the bending or 'elbowing' of the straw at the point of infestation. (Fig. 44).

DISTRIBUTION. Most prevalent in the eastern counties of England, but may be found all over the south and west and in Yorkshire. In Scotland it occurs from Cromarty to the Moray Firth, in Lothian, Perthshire, Haddington and Berwickshire.

CONTROL. Owing to the importance of the Hessian fly in America, an immense literature dealing with its control has arisen, but many of the methods advocated are not applicable to the pest in this country. It is important to destroy the pupae or 'flaxseeds' found in the screenings when threshing in-

Fig. 44. Wheat plants damaged by *Mayetiola destructor* Say. The Hessian Fly. 1, part of a wheat stem with sheath removed showing puparia above second joint. 2, characteristic elbowing of the stem caused by summer brood. About two thirds natural size. (After Criddle.)

fested wheat. Those pupae still in the stubble should be destroyed, by burning if possible, or by deep ploughing. Suitable crop rotation should be practised, clover is a safe crop to follow an infested cereal crop. Fertilisers to stimulate rapid growth are also useful. Flinty stiff-stemmed varieties of wheat are probably less liable to 'elbow', but McCulloch and Salmon (36) consider that soft wheat varieties have proved in general more resistant, a fact possibly due to physiological causes.

NATURAL ENEMIES. The following Hymenopterous parasites have been bred from the Hessian fly—in England:

Chalcididae
Merisus destructor Say.
Merisus intermedius Linn.
Micromelus subapterus Riley
Pleurotropis epigonus Walk.
Semiotellus nigripes Lind.
Eupelmus karschi Lind.

Chalcididae
Euryscapus saltator Lind.
Tetrastichus rileyi Lind.

Proctotrypidae
Platygaster herricki Riley
P. longicaudata Kieff.

in Italy (Leonardi):

Chalcididae
Baeotomus rufomaculatus Walk.
Ceraphron destructor Say.
Eupelmus allynii French
E. atropurpureus Dolm.
Holcaeus cecidomyidae Ashm.
Homoporus subapterus Riley
Isosoma hordeii Walk.
I. secalis How.
I. tritici Riley
Inastasius punctiger Först.

Chalcididae
Merisus destructor Say.
M. fulvipes Cress.
M. intermedius Lind.
Pteromalus forbesii D.T.
Tetrastichus productus Riley
T. rileyi Lind.

Proctotrypidae
Platygaster canestrinii Rond.

in Ukraine:

Chalcididae
Eupteromalus arvensis Kurd.
Merisus destructor Say.

Chalcididae
Meraporus crassicornis Kurd.
Trichacis remulus Walk.

in New Zealand:

Proctotrypidae
Platygaster minutus
Pleurotropus (Entedon) epigonus Walk.

in the United States of America:

Chalcididae
Calosoter metallicus Gah.
Centrodora speciosissimus Gir.
Eupelmnus saltrator Lind.
Eupelmus allynii French
Eutelus mayetiolae Gah.
Meraporus febriculosus Gir.
Merisus destructor Say.
Micromelus subapterus Riley

Chalcididae
Nemicromelus fulvipes Forbes

Proctotrypidae
Platygaster herricki Pack.
P. hiemalis Forbes
P. vernalis Myers
Polyscelis modestus Gah.
Pseudimerus mayetiolae.

Contarinia tritici Kirby. Wheat Midge

DESCRIPTION. *Adult.* Barnes (19) describes the female midge as follows: "Antennae, 2 + 12, two basal segments golden yellow, flagellar segments grey, first and second flagellar segments fused; each flagellar segment with basal cylindrical enlargement and short neck, each enlargement with two rings of applied circumfila joined by a longitudinal thread, the distal ring sometimes with very short loops, and a basal and distal ring of stout setae; enlargement of first flagellar segment $1\frac{1}{2}$–2 times as long as that of second, enlargement of third flagellar segment about $3\frac{1}{2}$ times as long as neck, which is slightly longer than broad, enlargement of tenth flagellar segment about $2\frac{1}{2}$ times as long as neck, which is about $2\frac{1}{2}$ times as long as broad; terminal segment with rounded cylindrical elongation slightly shorter than neck of penultimate segment and bearing minute setae, as on all enlargements but not necks. Palpi with four

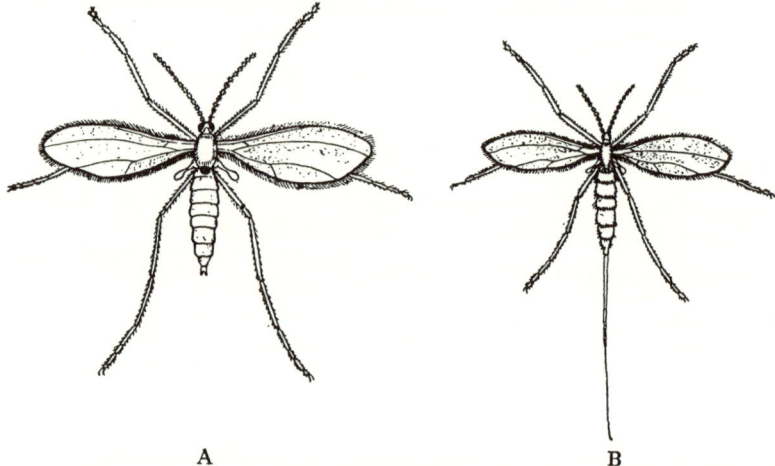

A   B

Fig. 45. Wheat Midges. *A*, *Sitodiplosis mosellana* Gehin. Female. *B*, *Contarinia tritici* Kirby. Female. Greatly enlarged. (From Barnes after Wagner.)

segments bearing long setae, basal segment small subglobular; second globular three times as large as first, third cylindrical, slightly longer and narrower than second; terminal segment slender, longer than third. Face golden yellow. Thorax pale brown yellowish. Wings transparent, iridescent; third vein interrupts costa at tip of wing, fifth vein forked. Abdomen golden yellow. Legs golden yellow with long pale hairs; claws simple, strongly curved, about as long as empodium. Ovipositor very long, retractile, aciculate, as long as abdomen and thorax when fully extended". Length 2 mm. (Fig. 45.)

Egg. Oblong, transparent, stalked or pedicellate, the stalk being longer than the rest of the egg. Golden yellow.

Larva. At first white then deepening in colour to lemon or golden yellow, tapering somewhat towards the head, transversely wrinkled. The anchor process, which lies ventrally near the head, is only slightly emarginate or notched, and in the anal extremity the posterior segment bears two very prominent rounded tubercles heavily chitinised, with a pair of smaller non-chitinised papillae armed with a small seta on the outside of the chitinised tubercles. The last pair of spiracles are also very conspicuous on the posterior margin of the eighth abdominal segment. These points are important to distinguish the larvae from those of the closely allied wheat midge *Sitodiplosis mosellana.* Length 2 mm.

Pupa. Reddish in colour, rather slender and pointed at each end.

LIFE HISTORY. In June the female midges appear and commence to oviposit in the wheat blossom, depositing the eggs in a pocket in the internal face of the glume, oviposition generally takes place during the evening, at 8–9 p.m., and 1–20 eggs may be laid at one time. The eggs hatch in about ten days and the larvae feed among the anthers and pollen at the top of the kernel. They do not, however, feed on the pollen, but on the juices drawn up into the ovary for the formation of the grain. (Cf. *S. mosellana.*) Infested ears may be recognised by the bluish colour of the base of the glume. In July the larvae are full fed and leave the ear to enter the ground for hibernation. They remain in the soil during the winter as larvae enclosed in a transparent case, only pupating about eight days before their emergence as adults in the following year. There is usually one brood in the year with occasionally a partial second. (Barnes in litt.) *C. tritici* chiefly attacks wheat, but occasionally rye and barley also, and probably couch grass (*Agropyron repens*).

DISTRIBUTION. Wheat midge was especially destructive in 1926 in the eastern counties, and severe infestations were noted in Cheshire, east Devon and Kent. Previously to this the midge was plentiful in various counties in the years 1916 and 1920.

CONTROL. Various measures of control have been suggested by different authorities, probably deep ploughing to bury the pupae so deeply as to prevent emergence, and suitable alternation of crops are the most effective. Burning the chaff to destroy the accompanying pupae is a doubtful measure, as Kieffer[33] considers that pupae remaining in the wheat when harvested are almost invariably parasitised, and such a measure would only destroy these beneficial insects. Later sowing of wheat so that it should flower after the oviposition period of the midge is also recommended. The varieties of wheat which suffer most are those which come into ear first, towards the end of June, there being possibly some connection between the stage of flowering of the wheat and the ability of the midge to insert its delicate ovipositor between the glumes. Wheats which come into ear later appear to suffer less.

NATURAL ENEMIES. Two larval parasites are recorded in Britain:

Proctotrypidae
  Leptacis tipulae Kirby
  Inostemma inserens.

Leonardi gives the following Hymenopterous parasites in Italy:

Ichneumonidae
  Ichneumon penetrans Smith

Proctotrypidae
  Inastasius inserens March
  I. punctiger Först.
  Leptacis tipulae Först.

Proctotrypidae
  Platygaster scutellaris Walk.
  Sactogaster pisi Först.
  Synopoeas muticus Thoms.

Chalcididae
  Pteromalus unicans Oliv.

Sitodiplosis mosellana Gehin. Wheat Midge

DESCRIPTION. *Adult.* "Female. Antennae, 2 + 12; two basal segments orange to carrot-red, flagellar segments grey; first and second flagellar segments fused; each flagellar segment with basal elongated enlargement with moderately long neck, each enlargement with applied circumfila and proximal and distal ring of long stout setae; enlargement of first flagellar segment not much longer than that of the second, enlargement of third flagellar segment about twice as long as neck, which is about three times as long as broad, enlargement of tenth flagellar segment about twice as long as neck, which is about four times as long as broad; terminal segment with rounded cylindrical elongation slightly shorter than the neck of penultimate segment and bearing small setae as on all enlargements but not necks. Palpi with four segments bearing long setae; proximal segment quadrate; second rectangular about twice as long as first; third cylindrical about as long as second; terminal segment nearly one third as long again as third and slightly swollen distally. Face orange yellow. Thorax pale brown dorsally otherwise yellowish. Wings dark transparent, bluish iridescent; third vein interrupts costa at tip of wing, fifth vein forked. Abdomen orange to carrot-red. Legs straw-coloured, with slightly darker hairs; claws simple, moderately curved about as long as empodium. Ovipositor moderately long, about one half the length of abdomen when fully extended, lamelliform, lateral lamellae about 2½ times

as long as broad, basal lamella small. Length 2–2·5 mm."
(Barnes(19).) (Fig. 45.) ˈ

Egg. Similar to that of *C. tritici*, but of a more orange or reddish yellow.

Larva. Orange-yellow, stout; head small rather long; antennae stout bi-articulate; anchor process bidentate, the teeth diverging, obliquely truncate, the shaft long, slender and tapering posteriorly; skin coarsely shagreened; posterior extremity roundly truncate and with two sub-median pairs of rather obtuse tubercles, the outer pair distinctly smaller. Length 2·5 mm. (Felt(28).)

Barnes differentiates this larva from that of *C. tritici* by its reddish rather than yellow colour, by the depth of the notch in the anchor process (see Fig. 50, 3), and by the presence, on the anal extremity of the larva, of four anal papillae slightly chitinised at their extremities.

Pupa. Orange-red, slender, rather pointed at the ends.

LIFE HISTORY. The larvae hibernate in the soil or chaff, the adults emerging in early June, particularly during a spell of hot sultry weather. The females oviposit when the ears of wheat are formed, at the tip of the husk covering the soft kernel. The larvae mature before the grain hardens, and drop to the ground preferably during rain or heavy dew; they penetrate about an inch below the ground where they pupate. In America rye is more heavily infested than wheat; the larvae may, however, attack wheat, rye, oats or barley. In wheat fields infestation may be detected by rubbing the heads in the hand, whereupon the larvae, if present, will drop out of the bracts. Infested rye can be recognised by a yellowish tinge showing through the bracts. Cool moist weather is said to be favourable to the larvae and to the activities of the adults by keeping the grain succulent; dry warm weather restricts the period during which the females can deposit their eggs.

DISTRIBUTION. It is probable that much damage previously attributed to *C. tritici* was due to this species. In Kent *S. mosellana* was especially injurious in 1926, 100 per cent. of the ears being attacked at Wye in one case. In 1927 this insect was again unusually abundant in east Devon.

CONTROL. Alteration in times of sowing and methods of cultivation seem to have little effect owing to the prolonged period of flight of this midge. Probably the best measures are good preparation of the ground and any means which will promote vigorous growth of the crop.

POINTS OF DIFFERENCE BETWEEN *C. TRITICI* AND *S. MOSELLANA*

As these two species of midges are so often confused, the following points of difference quoted from Barnes (19) will serve to differentiate them. In the egg, larva, and adult stage *S. mosellana* is orange to carrot-red, in the same stages *C. tritici* is golden yellow in colour. In the field, a further difference is the position of the larvae in the wheat kernel. The larvae of *S. mosellana* are to be found actually on the face of the kernel itself, while those of *C. tritici* are to be found in the brush among the anthers and pollen at the top of the kernel. (Fig. 46.) Morphological differences in the larvae are dealt with in the description of these stages in their respective species.

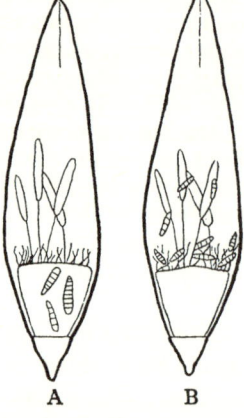

A   B

The following five species of midges are occasionally destructive to willows and osiers and are dealt with shortly:

Fig. 46. Wheat Midges; position of larvae in flower of wheat (diagrammatic) *A, Sitodiplosis mosellana*; *B, Contarinia tritici.* (After Barnes.)

Rhabdophaga heterobia H. Lw.
R. rosaria Lw.
R. salicis Schrank.
R. saliciperda Duf.
Dasyneura terminalis H. Lw.

Rhabdophaga heterobia H. Lw. Button-top Gall Midge

The adult midge is dusky brown or reddish, the under surface of the abdomen yellowish, the legs and wings with dark scales and hairs. The eggs are bright red, and the larva is orange-red and measures about 2·5 mm. in length. According to Barnes (20) there are two main broods in the year, the adults of the first brood emerging from April to July, and the second from July to September. The eggs are laid in the terminal rods of willows, especially *Salix triandra* and *S. cinerea*. Numbers of eggs are laid in one shoot and these hatch in about a week, the feeding of the larvae giving rise to a rather shapeless rosette or 'button' gall. When full fed the larvae either pupate in the gall or make their way to the exterior, fall to the ground and pupate in the soil. In addition to the production of the button-top gall, the midge has two alternative modes of attack. In one the eggs are deposited in the male catkins, which become deformed, the stamens and scales are thickened and the affected parts white and woolly in appearance. The second alternative is the attack on the lateral buds which become swollen and may contain up to 12 larvae. This is a destructive insect when it occurs in numbers as it attacks the varieties of willows most used for wicker work, stunting the rods and making them 'bush topped.' By the stoppage of end growth, side shoots arise from the lower lateral buds, which are of little use for basket-making. There are two main sets of button galls, one formed in the early part of the year near the ground, the other formed later, higher up the rods.

CONTROL. In October the bolts of rods should not be left in the fields, as this enables the larvae to leave the rods and enter the ground. The bolts should be removed at once to the pits or stacked upon a definite surface like concrete or cinders, where the emerging larvae could be dealt with by means of a suitable spray. All peelings should be burned. In the case of a slight attack, the collection and destruction of the galls, before the development of the insect is completed, are recommended. Barnes (*Ann. App. Biol.* XVII. No. 3, 1930), finds that certain varieties of osier are resistant to *R. heterobia,* and this may prove a good line of attack against the midge.

Rhabdophaga rosaria Lw.

Female with a grey hairy thorax and pink abdomen ending in a telescopic ovipositor. Under surface flesh colour with fine grey hairs. Legs, long, thin and grey; tarsi reddish. Wings grey with darker grey veins; halteres pale; wing expanse, 8 mm. The male is all grey, the thorax being darker than the abdomen and hairy. (Theobald (48).) The life history is similar to that of *R. heterobia* but the galls are more symmetrical and more rosette-shaped, they contain a single larva only instead of a number as in the preceding species.

NATURAL ENEMIES. The following Hymenopterous parasites are recorded in Italy (Leonardi): *Apanteles falcator* Marsh., *A. scabriusculus* Reinh., *Acoelius subfasciatus* Hal., *Bracon vitripennis* Ratz., *Hemiteles arcator* Grav., *Elasmus nudus* Först., *Encyrtus sitalces* Walk., *Torymus impar* Rond. and *Tridymus rosulorum* Ratz.

Rhabdophaga salicis Schrank

The adult female midge has a dark grey thorax with two yellow-ish lines; abdomen dark grey, paler, often red below; legs long and grey; wings reddish at base. The flies appear in May and oviposit in the previous year's twigs where the resulting orange larvae form woody galls. There may be 10–40 larvae in a gall. The winter is passed in the larval state within the gall, in the spring the larvae pupate *in situ*, the adults hatching in May.

NATURAL ENEMIES. Leonardi records the following Hymeno-ptera from *R. salicis* in Italy: *Acoelius subfasciatus* Hal., *Aphidius ambiguus* Hal., *Ceraphron clavatus* Ratz., *Encyrtus sitalces* Walk., *Hemiteles arcator* Grav., *Platygaster cecidomyiae* Bi. and *Tetrastichus capitatus* Ratz.

Rhabdophaga saliciperda Duf. The Shot Hole Borer

Theobald (48) describes this fly as having the head and thorax black with black hairs, the wings milky white with whitish hairs, with a wing expanse of about 4 mm. The pupa is yellow with two well-marked horns at the base of the antennae. The eggs are de-posited in May and June in the bark of willows, mostly of two years' growth, the eggs being laid in strings or clusters, the resulting larvae tunnelling in the bark. The presence of the larvae is shown by a swelling, often of a spindle shape, which later splits. The larvae continue to live in the wood till the following June when they pupate near the surface of the bark. Before hatching, the pupae work their way to the exterior by means of the projecting horns; after emergence of the adults the exit holes are left in the bark giving a 'shot hole' effect. As a control, smearing the affected places, recognisable by the swelling, with tar to prevent emergence of the adults is recommended.

NATURAL ENEMIES. Leonardi records the following Hymeno-ptera parasitising *R. saliciperda* in Italy: *Eurytoma nobbei* Mayr., *E. saliciperda* Mayr., *Platygaster cecidomyiae* Bi., *Pteromalus citrinus* Ratz., *P. diadema* Ratz. and *Torymus fuscipes* Boh.

Dasyneura terminalis H. Lw.

This midge, which is brownish black in colour, mainly attacks the crack willow (*Salix fragilis*) and the common willow (*S. alba*). The eggs are deposited in the apices of the shoots, which swell up to form a gall; there may be as many as 20 or 30 larvae in a gall.

Contarinia nasturtii Kieff. Swede Midge

DESCRIPTION. *Adult.* The following description is abbreviated and translated freely from Kieffer(33).

Male. Head, eyes black, broader at the top, contiguous, lower part of face, proboscis, and palps whitish; antennae 1·7 mm. long, 2 + 12, the segments simple with one whorl, first flagellar segment without stem, the succeeding segments globular with stems which are somewhat longer than the segments, the last segment with two

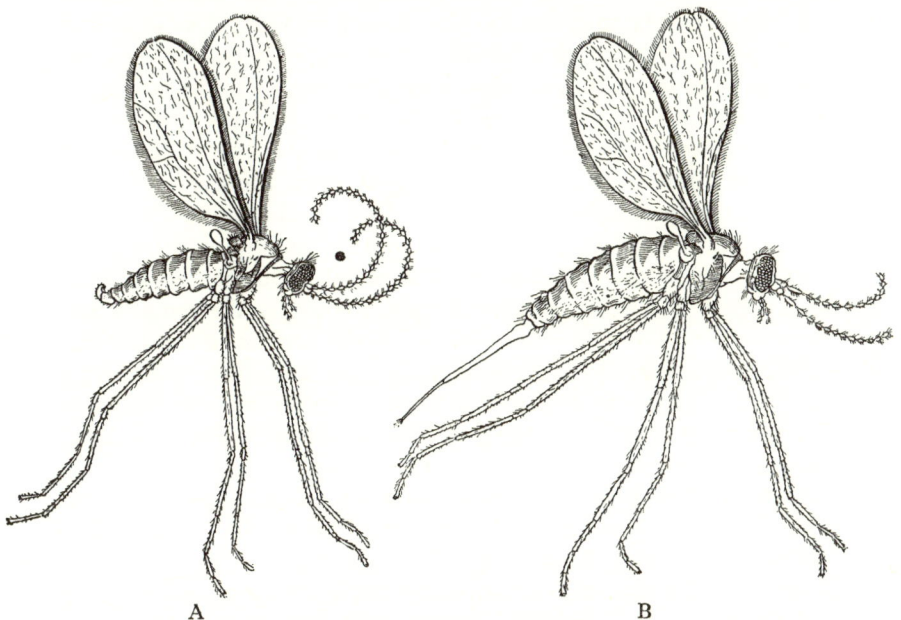

A   B

Fig. 47. *Contarinia nasturtii* Kieff. The Swede Midge. *A*, male; *B*, female. Greatly enlarged. (After Taylor.)

whorls; neck yellow, thorax lemon-yellow, prothorax and scutellum slate grey, on the latter are two shiny black furrows, running lengthwise in which are two rows of grey hairs, between the first and second coxa is an oblique patch which stretches sideways as far as the humerus. Wings 1·6 mm. in length, strongly iridescent, long and wedge-shaped at the base. Abdomen lemon-yellow with some indistinct transverse stripes. (Fig. 47.)

Female. Antennae 0·8 mm. long, 2 + 12, the first flagellar segment is constricted in the middle and is twice the length of the succeeding ones, which are cylindrical with short stems; ovipositor

whitish and capable of considerable extrusion, it is pointed like a
needle. Length 1·6 mm. (Fig. 47.)

Larva. Pale yellow in colour, legless, with the power of jumping.
Length 2–2·5 mm.

LIFE HISTORY. The bionomics of the swede midge have been
studied by Taylor (45), from whose account much of this information
is derived. There are usually three broods in the year, with occasionally a fourth in a hot summer, but they are not sharply defined
and much overlapping occurs. The flies of the first brood appear in
June and the succeeding broods until September. The eggs are
deposited in clusters or strings, sometimes on the upper surface of
the swede leaf, but more frequently on the upper surface of the
stalk or the younger leaves in the heart of the crown. The eggs
hatch in a few days and the larvae feed upon the leaf epidermis,
they do not mine below the surface. When full fed, in about three
weeks, the larvae descend to the ground and pupate in silken
cocoons below the soil surface. The pupal period in the summer
brood lasts 2–3 weeks; before the adult emerges, the pupa breaks
out of the cocoon and makes its way to the surface of the ground,
while projecting above the soil the pupal case splits and the fly
emerges. The larvae hibernate in a cocoon in the soil and pupate in
the spring, the adults emerging in June (Thomas (1946), *Ann. Appl.
Biol.* XXXIII, 77–81).

CULTIVATED HOST PLANTS. Mostly Crucifers, particularly the
swede, but *C. nasturtii* is also recorded from rape, radish, and
cabbage, and rarely from turnips.

WILD HOST PLANTS. Charlock (*Sinapis arvensis*), wild radish
(*Raphanus sativus*), marsh watercress (*Nasturtium palustre*) and
creeping watercress (*N. silvestre*).

SYMPTOMS OF ATTACK AND INJURY TO HOST PLANT. In the case of
the swede, after the larvae have fed for a few days on the young
tissues, the normal growth of the plant becomes altered, the extent
of this alteration depending on the age of the plant and the number
of larvae present. The stalks become swollen (Fig. 48) and, bending
sharply inwards across the top of the plant, press upon and compact
the terminal bud, while the leaf blades become crumpled. This
damage retards the general development of the plant and aids the
larvae to escape the effects of spraying. In addition to this 'crumpled
leaf' effect, the larvae cause the condition in swedes known as
'many-neck'. This injury is due to the June brood and is caused

by a number of larvae killing the growing point and thus inducing the formation of the many-necked condition. When savoys are attacked, 'blindness' or 'button-heart' is occasionally caused.

DISTRIBUTION. Widely distributed and common in many parts of Great Britain. In 1926 the midge was very plentiful in Wales, especially in the lowlands and coastal districts of Carnarvon, and was still prevalent the following year, severe losses being experienced in the southern counties and in Nottingham and Lindsey.

CONTROL. The attacks of the midge in fields sown in May are sometimes greater than those occurring in fields sown in June, though sowing date seems to have little effect. Heavy rain is probably harmful to the midge by washing the eggs off the plants

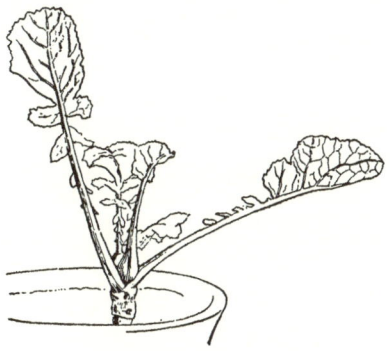

Fig. 48. Young swede plant attacked by the larvae of *C. nasturtii*. Note the swollen bases of the young leaves which constrict the growing point. (After Taylor.)

and interfering with the flight and oviposition of the adults. Taylor suggests a trap crop of one or two rows of swedes along the headlands of fields likely to be attacked. When the eggs have been laid in such a trap crop, and the midges have died off, the swedes should be lifted and destroyed together with the larvae they contain, without delay. Spraying is not recommended as being troublesome and ineffective. The numbers of this midge appear to be influenced by climatic factors, and Dry (26) considers that the progress made by the midge varies roughly with the average mean air temperature or with the numbers of hours of bright sunshine.

NATURAL ENEMIES. Dry records an Empid fly as attacking the adult midges and he has also reared some undetermined Procto-trypid parasites from the larvae.

Dasyneura brassicae Winn. The Pod Midge; Turnip and Cabbage Seed Midge

DESCRIPTION. *Adult.* Male. Antennae one fifth shorter than body, brownish black, 2 + 13, segments and stems of equal length with long whorls of hair, palps whitish, lower part of face brown with silvery white hairs, top of head brownish black, orbits with silvery white hair. Thorax black, lower surface brownish black with three small patches of silvery white hairs. Abdomen flesh-pink with black bands which are often absent, under surface pink with short silvery pubescence. Legs black with silvery white pubescence. Wings transparent, iridescent, costa black, veins black.

Female. Antennae not quite half the length of body, 2 + 13, with whorls of short hair, flagellar segments cylindrical without stems; body flesh-pink or red dorsally, with broad black bands of scaly hair; ovipositor yellowish white and capable of extrusion for a considerable distance. (Abbreviated from Winnertz(51).)

LIFE HISTORY. The winter is passed in the pupal condition in the soil, the adults appearing in early summer. The life history of *D. brassicae* seems bound up with that of *Ceuthorrhynchus assimilis*, the turnip seed weevil (p. 141). W. Speyer(44) thinks that the midges oviposit in the pods of Cruciferous plants in the punctures made by the weevil, as their ovipositors are considered to be too delicate to pierce the pod. Laboulbène, on the other hand, states that the female can pierce the pod. It seems certain, however, that serious attacks of the pod midge can be correlated with attacks of *C. assimilis*. The larvae are white and non-jumping, they live gregariously in the pods of Cruciferous plants, especially field cabbage, the larval stage being about four weeks. The chief symptom of attack is a premature yellowing of the pod. There are several broods in a year, up to six if the seed weevil is also present and weather conditions are favourable. Plants attacked include rape, turnip, radish, mustard, swedes and occasionally beet. It is sometimes a serious pest in Romney Marsh and Thanet, and it is common in Cambridgeshire, where the damage caused is known as 'bladder-pod'. No satisfactory control measures are known, deep ploughing would probably destroy the overwintering pupae.

The following parasites of *D. brassicae* are recorded from France and Germany: the Proctotrypid *Inostemma boscii* Jur., *Platygaster nigra* Nees., and the Chalcids *Pseudotorymus brassicae* Rusch. and *Tetrastichus brevicornis* Thoms.

The larvae of another species of midge are also occasionally present in the flowers of Cruciferous crops. This is *Dasyneura raphanistri* Kieff.; it is yellowish brown in colour and the larvae are white and non-jumping, living gregariously in the flowers. The life history is similar to that of the preceding *D. brassicae.*

Clover Midges

Amblyspatha ormerodi Kieff. The Red Clover Gall Gnat

DESCRIPTION. *Adult.* Brownish black, in the female the abdomen is dark red with transverse brown bands. Palps with four segments, the first two short. Antennae, in the male, of 14 joints of which the third is longer than the fourth, 4th–13th almost globular, a little longer than broad, distinctly shorter than their collar, 14th segment the shape of an obtuse cone. Length 1·5 mm.

Larva. Narrow, smooth, head short and brown, colour red, orange-red, or pinkish. There is a characteristic anchor process, consisting of a long and slender stem surmounted by a head twice as broad as long, rounded before and behind with the greatest length in the middle.

LIFE HISTORY. The female oviposits mostly on red clover, the eggs being probably laid on unopened buds or young leaves. The larvae feed in various places; in the tap root, at the apex of the plant, or in unopened buds. The eelworm *Tylenchus dipsaci* Syn. is often associated with the damage due to the larvae of this midge.

McDougall(37) describes the damage as follows: (1) Plant diseased at soil surface, at the junction of root and stem. (2) Side shoots decayed, shoots deformed. (3) Stipules of unexpanded leaves often dead and blackened, leaflets discoloured. (4) Primary root in smaller plant sometimes destroyed, in larger plants damaged at apex.

The disease usually appears when there is an abundant autumn growth in the plants after the corn is cut, the usual practice being to sow the clover with barley or oats. Wet weather after harvest appears to be favourable to the fly. Widely distributed in Great Britain, to which it seems to be confined. It chiefly attacks red clover (*Trifolium pratense*) but has been recorded from kidney vetch (*Anthyllis vulneraria* Linn.). McDougall states that he has received complaints from November to March from many counties

180 DIPTERA

in the British Isles, and Theobald reports it as frequently attacking red clover in Kent. Control measures include feeding off the clover to sheep, or cutting after the corn is harvested, and ploughing deeply crops which are badly infested.

Dasyneura trifolii F. Loew. The Clover Leaf Midge

DESCRIPTION. *Adult.* (After F. Loew(34a).) Male. Head small, brownish black, eyes large and black, bordered with pale hairs, palpi small and yellow; antennae 14–15 joints, 1 mm. long, dusky brown, each joint with two whorls of yellow hair. Thorax reddish brown with two small black longitudinal furrows. Halteres pale yellow. Length 1·3 mm.

Female. Darker coloured, antennae slightly shorter, 14–15 joints, ovipositor long, telescopic, yellowish brown, lighter towards end. Length 1·6 mm.

Larva. Light reddish yellow. Length 1·5–2 mm.

Pupa. Reddish brown, forepart dusky brown. There is a white silky cocoon which the pupa leaves before emergence. Length 1·2–1·5 mm.

LIFE HISTORY. The eggs are deposited on the leaves of white clover (*Trifolium repens*) before the leaflets have expanded, the eggs being generally deposited low down on the plant. The infested leaflets never expand but become brown and swollen; in this gall the larvae live and pupate, the last brood probably pupating in the soil for hibernation. Barnes(18) describes the gall as follows: "It consists of one or more of the leaflets remaining folded along the midrib, so that the upper surfaces are contiguous. It becomes hard, but rather fleshy and inflated towards the midrib, the inflated part is yellowish green or reddish. Usually only one of the three leaflets is affected, sometimes two".

This midge is widely distributed over Europe, the United States of America and Canada. It has been recorded from every county in Great Britain except Westmorland.

Dasyneura leguminicola Lintn. The Clover Seed Midge; The Clover Flower Midge

DESCRIPTION. *Adult.* (After Lintner.) Antennal segments vary in number from 13 to 17, the most common number in both sexes being 16, basal joint coloured and short, the next joint black, short and naked, the remaining joints with whorls of hair. Wings, with numerous short curved blackish hairs; halteres reddish yellow. Abdomen dark brown, marked on each segment dorsally with black hairs. Thorax black above, clothed with long hairs. The marked features are the genitalia, a pair of broad extended claspers in the male, and the long jointed ovipositor in the female.

Egg. Elliptical in shape, shell smooth and transparent, colour at first pale yellow later becoming orange. Length 0·3 mm.

Larva. Orange red or pinkish, surface of the skin granulated; there are thirteen body segments with nine pairs of respiratory tubercles situated respectively on segment 2 and segments 5–12 inclusive. All these tubercles are lateral in position excepting the first and last pairs which are dorsal and posterior on their respective segments. (Folsom (29).) The anchor process is large and serves to distinguish this larva from that of the foregoing clover leaf midge (*D. trifolii*). (Fig. 50.)

LIFE HISTORY. The adults appear at the time of flowering of the clover and the eggs are laid in the opening florets in green clover heads. The eggs are usually laid singly, but sometimes in clusters of 2–5, in one head of clover there may be 50 eggs. The larvae feed inside the corollas on the young ovules and prevent seed formation. The full-grown larvae wriggle out of the closed florets and fall to the ground, where pupation takes place just below the surface in oval cocoons composed of fine silk and particles of earth. The pupa is pale orange with brown eyes and a pair of short conical tubercles on the front of the head; this stage lasts about 10 days. There are probably two or three broods in the year. In Great Britain, red clover is mostly attacked but there are records of it also from white clover and alsike (*Trifolium hybridum* Linn.). According to Wehrle (50), in America, larvae from eggs laid in late June pupate in September or the following spring, those from eggs laid in the latter part of August or in September overwinter in the cocoon and pupate in the following spring, usually about the middle of May. Before the adult emerges, the pupa makes a hole in the cocoon and comes to the surface of the soil.

CONTROL. Infested seed may be subjected to gentle heat in the open air, or fumigated in an airtight container with chloroform or carbon bisulphide. Moznette (41) recommends some practice which will change the normal time of blooming of the clover. If both a hay and seed crop are desired, hay should be mown ten days or two weeks earlier in order to prevent the development of the larvae by the drying up of the food plant. This practice hastens the development of the second crop of clover heads, so that the second generation of midges have only a few green heads in which to oviposit. Infestation in the field may be recognised by the uneven blossoming of the clover heads.

NATURAL ENEMIES. The Proctotrypids *Inostemma leguminicolae* Fouts, *Platygaster leguminicolae* Fouts and *P. error* Fitch and the Chalcid *Tetrastichus carinatus* Forbes are recorded from this midge in North America. Wehrle(50) gives an account of the natural enemies of *D. leguminicola* in America.

Contarinia pisi Winn. Pea Midge

DESCRIPTION. *Adult.* Male. Antennae pale yellow; mouth parts and part of head yellowish; antennae darker at the apex, 24 segments, segmental hairs shiny. Thorax brownish yellow sometimes paler, under surface dusky. Abdomen yellowish with narrow dark bands and greyish hairs. Wings transparent, dusky with darker costa, the first long vein does not reach quite to the middle of the costa, the second reaches the apex of the wing where it is curved downwards, the third bends downwards, its branches somewhat pallid. Legs dusky with yellow bases and under side and base of the femora, some hairs on the femora and tibiae. Length 1·8–2 mm.

Female. Antennae of 12 segments; abdomen pale yellow with dark hairs, long extensile yellow ovipositor. Legs dark brown to blackish. (Theobald.) Length 2–2·5 mm.

Larva. White in colour with the power of jumping. Characteristic brown anchor process. Length 3 mm. (Fig. 49.)

LIFE HISTORY. There are probably two broods in the year, the winter being passed in the pupal state in the soil. There are two types of injury to the pea, the first of which has recently been described by Barnes(22). In this the female oviposits in the flower of the pea. The flowers containing the larvae, of which there may be a large number, can be recognised by their swollen appearance, particularly the bottom part of the sepals. The petals are crinkled and deformed while the sepals are of normal length. Damage by the pea thrips is similar (p. 24), but the flowers in this case are more shrivelled. The later-appearing flies of the first brood and those of the second brood oviposit in late June and July in the pods of the later varieties of peas. The writer has found these larvae in large numbers in July in the pods of maincrop peas, where they appeared to damage both the inner face of the pod and the pea itself (Fig. 49). Any number up to 100 larvae may occur in one pod. As to whether the fly oviposits in the flower or the pod seems to depend upon the time of flowering of the pea. Beans are also sometimes attacked. This insect is widely distributed throughout

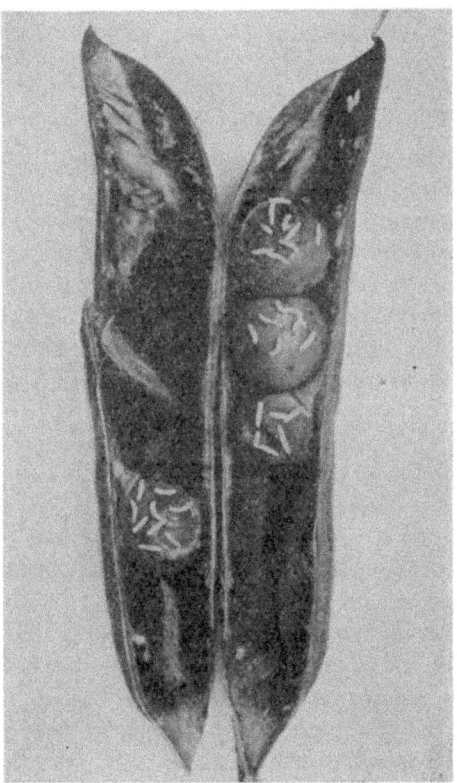

Fig. 49. Larvae of *Contarinia pisi* Winn.,
the pea midge, *in situ* in pea pod. × 1½.

Great Britain and over most of Europe; the writer has found it abundant occasionally in Cambridge and Theobald records it as plentiful in Kent. There is no good method of control for this midge, the land should be deeply dug after harvesting the crop and it may be treated with naphthalene at the rate of 3 cwt. to the acre and should be top worked only, in the spring.

Barnes records a midge which is predacious upon *C. pisi;* this is *Lestodiplosis pisi* Barnes (M.S.), of which the larvae are bright blood-red in colour (*Bull. Ent. Res.* XIX, 1928). In Austria the Proctotrypid *Sactogaster pisi* Först. is recorded as parasitising *C. pisi* upon *Pisum sativum,* and in the United States of America the Scelionid *Polymecus hopkinsi* Cwfd. and Brad. is recorded.

Grass Midges

The larvae of three species of Cecidomyidae live in the heads of meadow foxtail grass (*Alopecurus pratensis*), thus preventing the formation of seed, and these are dealt with shortly. According to Barnes(23), from whose work this information is derived, the three species are *Dasyneura alopecuri* Reut., *Stenodiplosis geniculati* Reut. and *Contarinia merceri* Barnes. The winter is passed in the larval stage, either in the fallen seed cases on the ground (*D. alopecuri* and *S. geniculati*) or below the soil surface (*C. merceri*). Pupation takes place in the spring, the adults emerging shortly afterwards. *D. alopecuri* has, as a rule, only one generation a year; the females of this species have red bodies. *S. geniculati* is two-brooded, the first emerging in April and May, and the second from June to August. The bodies of the female midges of the summer brood are bright red while those of the spring brood are more sombre. *C. merceri* has one brood a year as a rule, the flies emerging from May to July; occasionally, however, both this species and *D. alopecuri* have partial second broods. The larvae of *C. merceri* leave the empty seed cases in late summer and, springing into the air, fall to the ground, where they remain until the following spring. The adult females of this species have bright yellow bodies. Barnes differentiates the larvae of these three species thus: those of *D. alopecuri* are orange to brick-red in colour, those of *S. geniculati* pale buff to salmon-pink and those of *C. merceri* bright golden yellow. As a rule only one larva of *D. alopecuri* and *S. geniculati* occur in a floret, while up to fifteen individuals of *C. merceri* may be found in a single floret. The chief measure of control lies in the prevention of the grass from flowering until the majority of the midges have emerged. This may be achieved either by grazing or by rough scything the grass, and then allowing it to flower and seed in the normal way. Care must be taken that cutting or grazing is not continued too long or the grass will not flower and produce seeds.

Seed can be freed of midge by dry heating to a temperature of 59°–60° C. for 35 minutes or fumigation with carbon bisulphide in a sealed room (1 gr. carbon bisulphide to a litre of air). The Proctotrypid *Prosactogaster attenuata* Hal. is a common parasite of these three midges.

Contarinia humuli Tölg. (*Diplosis humuli* Theob.).
The Strig Maggot

This midge is occasionally a pest of hops and is dealt with briefly.

DESCRIPTION. *Adult.* Male and female pale yellow, flagellae and legs brown; the whorls of hair on each flagellar segment reach to about the middle of the next segment following, the stems of the second and third flagellar segment are somewhat longer than the basal segment, and the stems of the following segments increase in length. In the female the first flagellar segment is longer than the second and is constricted in the centre, the following segments are progressively shorter and the end segment has a shaft covered with fine hair. The ovipositor is long, hairless and striped, with a needle-like point and a few protruding hairs. Length 1·5 mm. (Abbreviated from Tölg.)

LIFE HISTORY. The winter is passed in the larval stage within a semi-transparent puparium in the soil. Pupation takes place the following year, the midges emerging in time to oviposit when the hops are in 'burr'. The white 'jumping' maggots tunnel into the strig and around the base of the bracts causing them to turn brown and fall; as many as 50 larvae may occur in a cone. The maggots are usually full grown by the middle of September, when they jump from the cone and fall to the ground in which they bury themselves about 2 in. below the surface and there pass the winter. This midge has been recorded from Kent round Canterbury, Maidstone and Gravesend and from Surrey, Hampshire and Worcestershire. It was first recorded by Miss Ormerod from Kent in 1882.

CONTROL. Theobald recommends a dressing of powdered naphthalene, applied in early autumn at the rate of 3 cwt. to the acre; or the folding of sheep in infested gardens during the winter.

In Fig. 50 are given a number of outline drawings of the 'anchor processes' of the various Cecidomyid larvae dealt with. These drawings are intended merely as an aid to the identification of the larvae and must not be regarded as exact representations of the anchor processes.

186

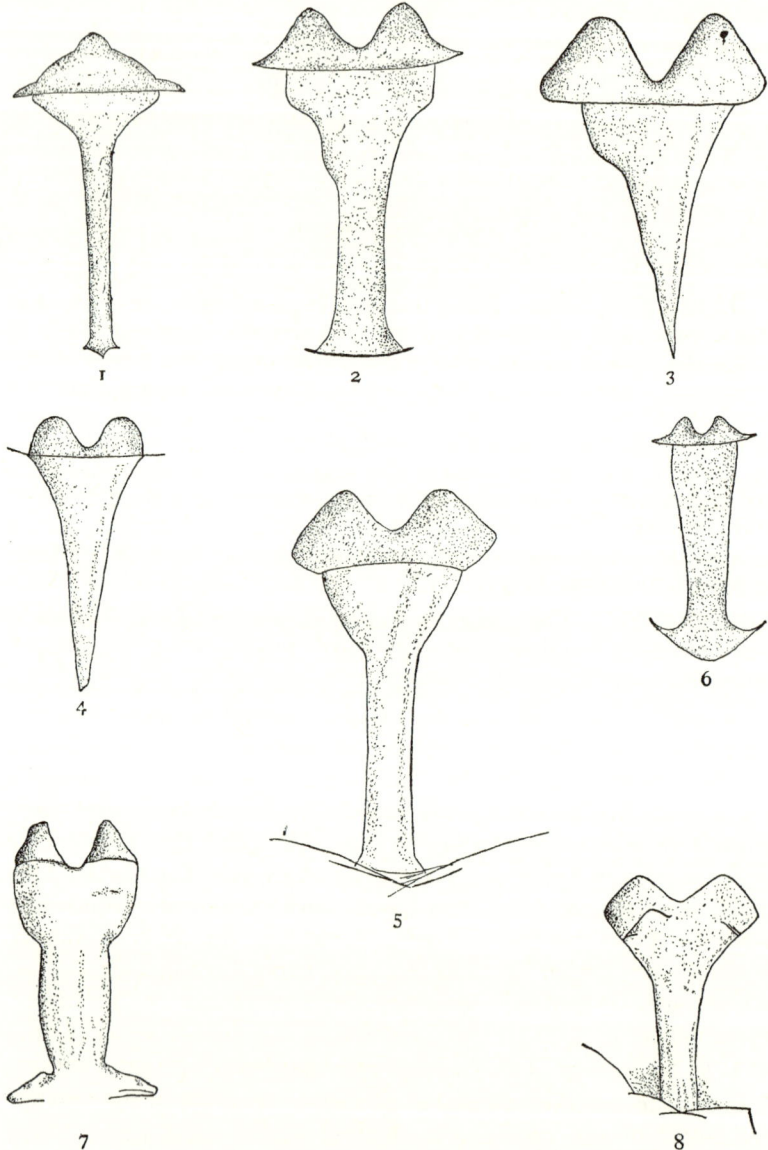

Fig. 50. 'Anchor Processes' or 'Breastbones' of various Cecidomyid larvae. Greatly enlarged. The shape of the anchor process is an aid to the identification of the species.   1, *Amblyspatha ormerodi*; 2, *Dasyneura leguminicola*; 3, *Sitodiplosis mosellana*; 4, *Clinodiplosis pisicola*; 5, *Contarinia pisi*; 6, *Dasyneura trifolii*; 7, *Mayetiola destructor*; 8, *Contarinia tritici*.

REFERENCES

DIPTERA
Tipulidae

(1) ALEXANDER, C. P. (1919–20). The crane flies of New York. *Agric. Exp. Station, Cornell Univ. Memoirs*, Nos. 25, 38.

(2) BODENHEIMER, FRITZ (1923). Beiträge zur Kenntnis von *T. oleracea* L. *Zeitschr. angew. Entom.* IX.

(3) BRITTEN, H. (1926). *The North Western Naturalist.* I, No. 2.

(4) DEL GUERCIO, G. (1913). La Tipule ed i Tafani Nocivi. *Redia*, IX.

(5) GASOW, H. (1926). *Mitteil. Dtsch. Landw. Ges.* XLI.

(6) Ministry of Agriculture and Fisheries, London. Leaflet 11.

(7) MORRISON, T. A. (1925). Species determination of two common crane flies, *Tipula paludosa* and *T. oleracea*. *Proc. Roy. Phys. Soc. Edin.* XXI, Part 1.

(8) OLDHAM, J. N. (1928). On the final larval instar of *Tipula paludosa* Meig. and *T. oleracea* Meig. *Proc. Roy. Phys. Soc Edin.* XXI, Part 5.

(9) RENNIE, JOHN (1916–17). On the biology and economic significance of *Tipula paludosa* Meig. Parts I and II. *Ann. App. Biol.* II, No. 4; III, Nos. 2, 3.

(10) RENNIE, JOHN (1927). Crane fly grubs and the oat crop. *Scott. Journ. Agric.* X.

(11) RENNIE, JOHN (1926). A Mermithid parasite of *T. paludosa*. *Proc. Roy. Phys. Soc. Edin.* XXI, Part 1.

(12) THEOBALD, F. V. (1903). *First Report on Economic Zoology*. British Museum (Nat. Hist.).

Tabanidae

(13) AUSTEN, E. E. (1906). *British Blood-sucking Flies*. British Museum (Nat. Hist.).

(14) CAMERON, A. E. (1926). Bionomics of the Tabanidae (Diptera) of the Canadian Prairie. Part I. *Bull. Entom. Res.* XVII.

(15) HADWEN, S. (1923). Insects affecting livestock. *Dom. of Canada, Dept. Agric. Entom. Br. Bull.* No. 29.

(16) PARMAN, D. C. (1928). *Experimental dissemination of the Tabanid egg parasite*, Phanurus emersoni *Girault. and biological notes on the species*. U.S. Dept. Agric. Washington, D.C. Circ. 18.

(17) WEBB, J. L. and WELLS, R. W. (1924). Horseflies: Biologies and Relation to Western Agriculture. *U.S. Dept. Agric. Bull.* No. 1218.

Cecidomyidae

(18) BARNES, H. F. (1927). Material for a monograph of the British Cecidomyidae or gall midges. *Journ. S.E. Agric. Coll. Wye, Kent*, No. 24.

(19) BARNES, H. F. (1928). Wheat blossom midges. *Bull. Entom. Res.* XVIII.

(20) BARNES, H. F. (1929). Button-top of willows. *Journ. Min. Agric.* XXXVI.

(21) BARNES, H. F. (1926). The gall midges of vegetables and market garden crops. *Journ. Roy. Hort. Soc.* LI.

(22) BARNES, H. F. (1927). New damage to peas by the pea midge. *Journ. Min. Agric.* XXXIV, No. 2.

(23) BARNES, H. F. (1930). On the biology of the gall midges attacking meadow foxtail grass. *Ann. App. Biol.* XVII, No. 2.

(23A) BARNES, H. F. (1946). *Gall midges of economic importance*, Vols. I, II. Crosby, Lockwood and Sons, Ltd.

(24) COLLIN, J. E. (1904). A list of the British Cecidomyidae arranged according to the views of recent authors. *Entom. Mon. Mag.* XV, No. 2.

188   DIPTERA

(25) CRIDDLE, N. (1915). The Hessian fly and the western wheat-stem sawfly in Manitoba, Saskatchewan and Alberta. *Dom. of Canada, Dept. Agric. Entom. Br. Bull.* No. 11.

(26) DRY, F. W. (1915). An attempt to measure the local and seasonal abundance of the swede midge in parts of Yorkshire over the years 1912–14. *Ann. App. Biol.* 11.

(27) ENOCK, F. (1888). Parasites of the Hessian fly. *Entomologist*, XXI.

(28) FELT, E. P. (1920). Thirty-fourth report of the State Entomologist on injurious and other insects of the State of New York, 1918. *New York State Museum Bull.* Nos. 231, 232.

(29) FOLSOM, J. W. (1909). The insect pests of clover and alfalfa. *Twenty-fifth Report State Entomologist of the State of Illinois.*

(30) FRYER, J. C. F. (1917). Insect pests of the basket willow. *Journ. Board Agric.* XXIV.

(31) HASEMAN, L. (1916). An investigation of the supposed immunity of some varieties of wheat to the attack of the Hessian fly. *Journ. Econ. Entom.* IX.

(32) JENKINS, J. R. W. (1926). Notes on clover midges. *Welsh Journ. Agric.* II.

(33) KIEFFER, J. J. (1888). *Ent. Nach.* XIV, 263.

(34) LEONARDI, G. (1922–25). *Insetti Dannosi e Loro Parassiti in Italia fino al* 1911.

(34a) LOEW, F. (1874). *Ver. Zool.-bot. Ges. Wien*, XXIV, 143.

(35) MCCULLOCH, J. W. (1923). Resistance of wheat to the Hessian fly. A progress report. *Journ. Econ. Entom.* XVI.

(36) MCCULLOCH, J. W. and SALMON, S. C. (1918). Relation of kinds and varieties of grain to Hessian fly injury. *Journ. Agric. Res.* XII.

(37) MCDOUGALL, R. S. (1913). The red clover gall gnat, *Amblyspatha ormerodi* sp. nov. Kieff. *Journ. Board Agric.* XX.

(38) MCDOUGALL, R. S. (1905–6). Osier midges. *Journ. Board Agric.* XII.

(39) MEADE, R. H. (1887). The Hessian fly in Great Britain. *Entomologist*, XX.

(40) Ministry of Agriculture and Fisheries, London. Leaflet No. 125.

(41) MOZNETTE, G. F. (1917). Three insects affecting clover seed production. *Oregon Agric. College, Corvallis, Bull.* No. 203.

(42) ORMEROD, E. A. (1891). *Reports of injurious insects for* 1890.

(43) SPEYER, W. (1921). Beiträge zur Biologie der Kohlschotenmücke (*Dasyneura brassicae* Winn.). *Mitt. biol. Reichsanst. Land- u. Forstw.* No. 21. 208–217.

(44) SPEYER, W. (1923). Kohlschotenrüssler (*Ceuthorrhynchus assimilis* Payk.), Kohlschotenmücke (*Dasyneura brassicae* Winn.) und ihre Parasiten. *Arb. biol. Reichsanst. Land- u. Forstw.* XII. 79–108.

(45) TAYLOR, T. H. (1912). *Cabbage-top in swedes.* Univ. Leeds and Yorks Council Agric. Educ. No. 82.

(46) THEOBALD, F. V. (1913). *Report on Economic Zoology.*

(47) THEOBALD, F. V. (1906–7). *Report on Economic Zoology.*

(48) THEOBALD, F. V. (1913). Osier midges. *Journ. S.E. Agric. Coll. Wye, Kent,* No. 22.

(49) WALTON, C. L. (1927). The swede midge in North Wales. *Journ. Min. Agric.* XXXIV.

(50) WEHRLE, L. P. (1929). The clover-flower midge (*Dasyneura leguminicola* Lintn.). *Agric. Exp. Station, Cornell Univ. Bull.* No. 481.

(51) WINNERTZ, J. (1853). Beiträge zur Monographie der Gallmuecken. *Linnaea Ent.* Berlin, VIII, 154–322.

CHAPTER XII

DIPTERA (*contd.*)

Family *Syrphidae*

Mostly brightly coloured flies of moderate to large size; in the wings there is a second margin parallel with the outer wing margin formed by certain other veins, and also a false vein or 'vena spuria' between the radius and median. The group is an important one owing to the aphis-eating habits of certain species. Three members of this family attack narcissus bulbs and occasionally onions, parsnips and potato, and are in consequence shortly dealt with here.

Merodon equestris F. The Large Bulb Fly

DESCRIPTION. *Adult.* Large hairy flies of bee-like aspect.

Male. Head, eyes brown, with short yellowish pubescence, contiguous for about half the length of the frons, ocelli brown, distinct. Face with bright shining pubescence, antennae dark brown to black. Thorax shining black with thick pubescence which varies greatly in colour from all tawny to tawny on first thoracic segment with black on remainder. Abdomen shining black with long thick pubescence varying in colour from all black to black and tawny or all tawny. Legs black; hind femora slightly dilated; hind tibiae dilated with a spur at the tip. Wings clear, without markings, veins blackish, squamae large with a fringe of long hairs, blackish or tawny. Halteres small, concealed.

Female. Eyes not contiguous, frons widening. Pubescence on thorax usually all black. Length 10–12 mm.

Egg. Oval, chalky white. Length 1·6 mm.

Larva. The mature larva is yellowish in colour, robust, somewhat rounded, segmentation distinct. Anterior spiracles brown, posterior spiracles black, somewhat raised. On each side is a short spine. Length 12–18 mm.

Puparium. Light to dark brown in colour. The anterior spiracles remain as a pair of horn-like processes, and the posterior spiracles as a black projection. Length 12 mm.

LIFE HISTORY. There is one generation per annum, the eggs are deposited singly in the soil or on the base of the plant, hatching in 6–15 days. The larva enters the bulb between the scales where it feeds; there is usually only one larva in a bulb. Pupation takes place in March or April, usually in the soil but occasionally also in the bulb. The flies are on the wing from May to July. Host plants are mostly narcissus bulbs; but the insect is also recorded from *Amaryllis*, *Galtonia*, etc. Theobald has found it attacking the wild hyacinth (*Scilla nutans*). Bulbs attacked by *M. equestris* are usually worthless.

Eumerus tuberculatus Rond. Lesser Bulb Fly

DESCRIPTION. *Adult.* This insect is very similar to *Eumerus strigatus* of which a fuller description is given. Collin(2) points out the following differences: hind femora with a slight rounded projection at the extreme base beneath basal joint of hind tarsi, also with a rounded laterally compressed projection at base beneath, somewhat hidden by the yellow pubescence. The chief difference lies in the appearance of the male genitalia. Length 5·5–7 mm.

Egg. White, shining, oval, sculptured longitudinally and tapering to one end.

Larva. The descriptions of this and the succeeding stage are given according to Hodson(5). "Subcylindrical, greyish white, anterior spiracles dorsal and widely separated. Posterior pair fused and prominent, chestnut brown in colour. Integument clothed with minute brown scattered spines, the segments bearing rows of small tubercles carrying larger spines. On the last pseudo-segment on each side of the posterior spiracular process and slightly ventral is a large tubercle. Slightly dorsal to this and on the penultimate pseudo-segment is a small twin tubercle, each twin terminating in a cluster of curved spines. More dorsal still and on the same pseudo-segment is a third tubercle intermediate in size." Length 7–9 mm.

Puparium. Somewhat pear-shaped, integument tough, yellowish white, darkening later. Anterior spiracles not visible at first, but extruded later. Posterior spiracular process prominent. Posterior tubercles present and skin armature similar to that of the larva but less prominent. Length 6–7 mm.

LIFE HISTORY. The winter is passed in the larval stage inside the bulb, pupation taking place in spring in the soil close to the bulb. The adults hatch in May and continue to appear till June. The eggs are deposited either on the sides of the bulb or in the soil adjoining. The eggs hatch in about three days, and the young larvae make for the nearest bulb, usually entering at the neck. According to Hodson, eggs may be deposited on healthy as well as already diseased bulbs, and consequently both are open to attack. When full grown the larvae, with the exception of those overwintering, usually pupate in the bulb itself. There may be one or two generations a year, with an occasional partial third.

Eumerus strigatus Flyn.

DESCRIPTION. *Adult.* Male. Head, eyes touching slightly in front of head, vertex shining, blackish, almost triangular. Antennae with third joint dark and silvery pubescence on its inner face. Thorax coppery, shining, two faint whitish lines on dorsum, pubescence yellowish, scutellum similar. Abdomen dull black with three white crescents. Wings brownish at base squamae glassy, pale, fringed. Halteres inconspicuous. Legs dark, with tibiae brownish; hind femora dilated; hind tibiae curved.

Female. Frons wider, eyes not contiguous, pubescence on frons whitish, hind tibiae less conspicuously curved, lighter in colour. Length 5–7 mm.

Larva. Very similar to that of the preceding *E. tuberculatus*, but the skin armature is slightly more pronounced.

DISTRIBUTION. Widely distributed over Great Britain, *Eumerus* spp. are particularly injurious in the Wisbech district. In 1927 *M. equestris* was unusually abundant in Cornwall and it appears to be increasing in that district. Of the two species of *Eumerus*, *E. tuberculatus* is probably the more abundant throughout the country.

CONTROL OF NARCISSUS FLIES. The hot water treatment is recommended for bulbs infested with the larvae, immersion for about three hours at a temperature of 110° F. will destroy the insects. Owing to the fact that the hot water treatment sometimes induces unsatisfactory flowering under forced conditions, Cole(1) recommends fumigation with calcium cyanide at the rate of 12 oz. per 100 cubic feet for four hours at a temperature ranging between 60 and 90° F. Hodson recommends fumigation with paradichlorbenzene as follows: 4 oz. of paradichlorbenzene per cubic foot for 120 hours, or 2 oz. per cubic foot for one week. The bulbs should be placed in airtight containers in layers about 4 in. thick, with crystals of the fumigant spread on the bottom of the container and covered with sacking to prevent contact with the bulbs. Fumigation with carbon bisulphide at the rate of 4–10 lb. per 1000 cu. ft. has not proved efficacious. The same authority advises as preventive measures that the lifted bulbs should be placed to dry and ripen under cover of a shed rather than in the open. Decoy traps of damaged bulbs may be allowed to become infested and then burned; they should not be exposed for a longer time than about six weeks, roughly the duration of the life cycle of the fly. In America, Wilcox and Mote (8) recommend application of a solution of corrosive sublimate, 1 oz. to 10 gallons of water. Five applications are made altogether, the solution being poured round the plants.

Family *Oestridae*. 'Bot Flies'; 'Warble Flies'

Large, hairy, bee-like flies, heads rather large, antennae short and inconspicuous, mouthparts vestigial. Larvae endo-parasitic in mammals.

Gastrophilus intestinalis de G. (*equi* F.).
The Common Horse Bot Fly

DESCRIPTION. *Adult*. Female. Head as broad as thorax; eyes black, bare; ocelli black, distinct, with forwardly directed bristles. Frons with light and dark orange markings; antennae deeply sunk, each one in a pit separated by a ridge, bases of antennae with orange-coloured bristles, third joint brownish with a paler mark on its inner surface. Surface of antennal cavity with whitish pubescence. Arista long, dark-coloured. Lower part of face with silvery pubescence which is brightly shining in life. Thorax brownish with brown pubescence usually darker in the centre and lighter at the sides; there is a patch of long flattened hairs at the juncture of meso- and meta-thorax; scutellum with a ridge of dense erect bristles, darker in the centre and giving the appearance of two tufts. Abdomen brownish with short yellow pubescence and some darker markings at the junction of the segments. In the female the abdomen is characteristically elongated, the genitalia conspicuous and darker in colour. Legs pale yellow with short black bristles and some short yellow hairs. Wings grey, veins brownish; across the centre of the wing runs a faint smoky band, somewhat S-shaped, anterior to this near the wing margin lies a single spot, while two other similar spots lie close together at the apex of the wing. Length 12–16 mm. (Frontispiece 2.)

Male. Similar, genitalia inconspicuous; there is a large triangular prolongation in the hind trochanters; abdomen rounded not elongate.

Egg. Pale ochreous to whitish, tapering to the end which is attached to the hair of the host. Boat-shaped, flatly oval, the free end of the egg rising from the hair is sharply truncate. The egg is attached to the hair by two sub-parallel lip-like flanges about two thirds of the length of the egg. It is somewhat flattened laterally, and marked by faint striations running at right angles to the long axis of the egg. (Fig. 51, 1.) Length 1·25 mm.

Larva. According to Hadwen and Cameron (16) the first-stage larva possesses 13 segments. The posterior spiracles are borne on the distal ends of two cylindrical processes arising from the last abdominal segment. The full-grown larva is yellowish with a pinkish tinge, it is barrel-shaped with eight segmental rows of

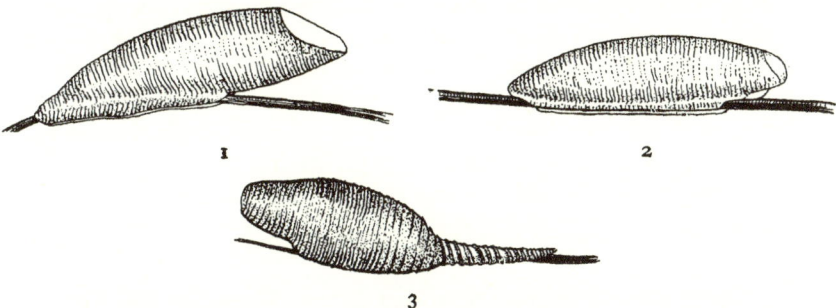

Fig. 51. Eggs of *Gastrophilus* spp. The Horse Bot Flies. 1, Egg of *Gastrophilus intestinalis* de G. The Common Bot Fly. The eggs may be found on almost any part of the body, but especially on the fore legs; colour yellowish white. 2, Egg of *G. nasalis* Linn. The Throat Bot Fly. The eggs are attached principally to the hairs between the jaws; colour yellowish white. 3, Egg of *G. haemorrhoidalis* Linn. The Nose Bot Fly. The eggs are attached to the lips of horses; colour almost black. All greatly enlarged. (After Hadwen and Cameron.)

spines, two girdles of spines on the segments, the spines in the second row smaller than those in the first; in the posterior spiracles the stigma occurs as a single plate which appears to be the fusion of the usual two stigmal plates; the openings are by two sets of three sinuous slits. There is a pair of mouth hooks present. It is the largest of the three species here described. (Fig. 52.)

Puparium. Jet black, smooth, shining, the double girdles of segmental larval spines plainly visible. Each double row is some distance from its neighbour. Length 12–14 mm.

LIFE HISTORY. The eggs of *G. intestinalis* are deposited mostly on the anterior surface of the fore legs either singly or in clusters. The writer has observed this fly ovipositing on horses and the favourite position was always low down on the inner face of the fore legs. Eggs are also deposited on the shoulders, along the line of the belly and high up on the hind legs. The fly hovers about the horse almost continuously with the body held nearly vertical while the long pointed abdomen cements the eggs to the hairs, one egg on a hair. Hatching of the eggs of this species is brought about by friction and this is supplied by the rubbing of the horse's lips. The larvae are ready to hatch in about seven days, but the eggs seem capable of remaining viable for many weeks. It does not seem quite clear how the larvae of *G. intestinalis* reach the stomach of the horse. Hadwen and Cameron consider that once on the animal's lips, they burrow under the tongue but their subsequent journey to the stomach has not been traced. The larvae live for about 9–10 months attached by their mouth-hooks to the pyloric end of the stomach. When full fed they lose their hold and pass out with the faeces. Pupation takes place a short distance under the surface of the soil, the pupal period lasting 20–70 days according to the temperature.

Gastrophilus haemorrhoidalis Linn. The Nose Bot Fly

DESCRIPTION. *Adult.* (Shortened after Lundbeck[20].) Male. Head, frons as broad as eye, orbit and frons yellowish, face yellowish white; antennae yellow at base, third joint darker, arista dark. Thorax black, somewhat shining, clothed with dense yellow pile; behind the suture a transverse band of black pile. Abdomen black shining, hind margins to segments reddish; first and second segments with long yellowish white hairs; third and part of fourth with short black hairs, apex with long red hairs. Legs yellow; femora blackish above. Wings yellowish; veins pale yellow to brown.

Female. Similar, frons a little broader, abdomen more black haired, sixth and seventh abdominal segments elongate, cylindrical. (Frontispiece, 3.)

Egg. Brownish black in colour with a grooved stalk, by which it is attached to the hairs. Longer than the eggs of the other two species. (Fig. 51, 3.)

Larva. The smallest larva of the three species. There are 13 body segments, colour pinkish until nearly mature when it becomes greenish. Two girdles of spines on each segment, the spines approximately of the same size. (Fig. 52.)

Puparium. Black, very similar to the foregoing species, larval spines less unequal in size.

Life history. The eggs are deposited singly on the hairs surrounding the lips of the horse, especially the lower lip. The eggs of
G. haemorrhoidalis are able to hatch without friction, and it is
thought that the young larvae penetrate into the stomach of the
horse of their own accord; hatching takes place within 5–18 days.
The larvae of this species first attach themselves to the wall of the
stomach, later passing to the rectum where they become re-
attached. Ultimately, before their final exit, the larvae attach
themselves for 2–3 days close to the anus and protrude to the
exterior. They do not appear to pass out with the faeces but become
detached at any time.

Fig. 52. Full-grown larvae of the three species of Horse Bot Flies. At left
G. haemorrhoidalis Linn., the Nose Bot Fly; in centre *G. intestinalis* de G., the
Common Bot Fly; at right *G. nasalis* Linn., the Throat Bot Fly. All much
enlarged. (After Bishopp and Dove.)

Gastrophilus nasalis Linn. Throat Bot Fly

Description. *Adult. Male.* Head, frons narrower than eye,
covered with short reddish pubescence, lower part of face with
longer orange-yellow pubescence; antennae brownish, arista black.
Thorax brownish black with dense shining orange pubescence,
scutellum sometimes darker. Abdomen black, shining, greyish
pubescence on basal segments, black on third and fourth segments,
while the apex bears bright orange hairs. Legs, femora black,
tibiae and tarsi yellow. Wings greyish, clear, veins yellow, no
markings.

196   DIPTERA

Female. Frons wider than eye; pubescence on the abdomen differs from that of the male, basal segments greyish, third segment black, apex of abdomen with sparse grey hairs, sixth and seventh segments drawn out ending in the genitalia. Length 12–13 mm. (Frontispiece, 1.)

Egg. Yellowish white in colour, attached to the hair of the host by two flanges which run the whole length of the egg instead of two thirds as in *G. intestinalis.* (Fig. 51, 2.)

Larva. Hadwen (15) describes the young larva of this species as having 12 body segments, and the only one bearing slender elongate hairs. It differs from the larvae of the other two species in having the posterior spiracles sessile instead of upon cylindrical processes. The full grown larvae are dirty white to yellowish in colour and rather broader than the two preceding species. There is only one girdle of spines on the segments. (Fig. 52.)

Puparium. Dull black, segmental girdles of larval spines single and close together. Length 10 mm.

LIFE HISTORY. The eggs are laid on the hairs beneath the jaws and occasionally on the shoulders, they hatch without friction. Hadwen (15) considers that the young larvae bore their way in through the skin, as he has seen small lesions on the skin of horses underlying the empty egg shells. The larvae remain in the stomach until the following spring when they pass out with the faeces.

DISTRIBUTION OF BOT FLIES. Very common and widely distributed throughout the British Isles, being perhaps rather more numerous further south.

INJURIES TO THE ANIMAL HOST AND METHODS OF TREATMENT. A good deal of the damage to horses by these insects is secondary, arising from the fear and annoyance caused by the ovipositing flies. At the same time the presence of large numbers of bot larvae in the stomach must be detrimental to health, as they cause considerable inflammation which interferes with digestion. Preventive measures consist in the provision of fly shelters such as darkened sheds, and the destruction of the eggs. This latter can be effected by clipping the hairs or singeing, but a more efficient method is to rub the affected parts with a cloth dipped in a 2 per cent. solution of any standard coal-tar creosote dip. For medicinal treatment of infected horses, carbon bisulphide is usually given. Hall and Avery (17) recommend three doses of 3 drachms each, at intervals of one hour, two doses of 4 drachms each at intervals of two hours, or a single dose of 6 drachms. The use of linseed oil as a purgative before dosing reduces the effectiveness of carbon bisulphide against bots. Carbon bisulphide must be given to horses with care and it is advisable for a veterinary surgeon to administer the drug.

The following characteristics of the three species of *Gastrophilus*, quoted from Lundbeck [20], are given to distinguish between these flies.

Adults. G. intestinalis. Wings with posterior cross vein, and dark transverse band and a spot at apex, hind trochanters in male with a large triangular prolongation, in female with a smaller one; hind femora suddenly thin at base.

G. haemorrhoidalis. Wings without markings, trochanters and hind femora simple. Posterior cross vein considerably behind medial cross vein.

G. nasalis. Wings without markings, trochanters and hind femora simple. Posterior cross vein a little before or just below medial cross vein.

Larvae. G. intestinalis. Two girdles of spines on segments; spines in the second row of the girdles smaller than those in the first.

G. haemorrhoidalis. Two girdles of spines on segments. Spines in the two rows less unequal in size.

G. nasalis. Only one girdle of spines on the segments.

NATURAL ENEMIES. Very little is known of the natural enemies of the Oestridae, and there appear to be no records of their parasites in Great Britain. In France Roubaud has induced the Chalcid *Nasonia brevicornis* Ashm. to parasitise *Gastrophilus intestinalis* under experimental conditions.

Hypoderma lineatum Vill. Ox Warble Fly; Heel Fly

DESCRIPTION. *Adult.* Male. General colour black with bands of yellowish and orange hair. Head covered with yellowish white hairs, frons and orbits brownish, paler than in the allied *H. bovis.* Thorax covered with black and white hairs, and there are four distinct longitudinal lines; scutellum with whitish hairs at sides. Abdomen with blackish hair on basal segment, black on median segment, terminal segments with hairs *orange-red.* Wings slightly fuscous, veins nearly black, alulae white. Legs, femora black, tibiae and tarsi brown.

Female. Frons much broader, abdomen more pointed with fewer orange hairs. Length 11–12 mm. (Fig. 53, *g.*)

Egg. The egg is dull yellowish white with a smooth shining surface, ovoid in shape, slightly larger at the base than in the middle.

It is attached to the hair by an oval clasp, the unattached end has a ridge across it along which the egg splits on hatching. Length 0·76 mm. (Fig. 53, *e*.)

Larva. When first hatched the larvae are glassy in appearance, creamy white in colour and covered with spines, the anal segment bearing spines of three types. According to Laake (19) there are five stages in the development of the larvae, the fifth stage or mature larva measures 16–26 mm. in length. The surface is rugged with a heavy spiny armature, and the posterior spiracles are somewhat kidney-shaped with radiating furrows, the last abdominal segment is without spines. (Fig. 53, *f*.)

Puparium. This retains many of the larval characters but is smaller, and becomes almost black in colour.

LIFE HISTORY. The adult flies appear in May and June with a period of maximum emergence in mid-July. Pairing takes place soon after hatching and, without feeding, the females commence to oviposit. The eggs of *H. lineatum* are attached to the hairs of the cow in rows by means of a clamp and with the addition of some adhesive substance; they are usually about twelve in number. The eggs are deposited mainly on the heels of standing cattle, but also on the hocks and occasionally on the sides and shoulders. (Fig. 54.) Hatching takes place in 3–6 days; the actual method of entry of the larvae into the host was for long a matter of conjecture, and the erroneous idea that the animal licked the eggs was accepted for many years. The work of Carpenter (13), Bishopp (11) and others has now demonstrated, however, that the larvae burrow directly through the skin and thus enter the body. On hatching, the young larvae make their way down the hair which bore the eggs, and after feeling about with the mouth parts begin to burrow into the skin at the base of the hair. It is probable that after the first larva has effected an entry, the others enter the host through the same hole. The larvae then apparently work upwards in the connective tissue and appear in the chest and abdominal cavity about two months later. Although many larvae enter the connective tissue of the gullet, it appears that some may not enter that organ. The next migration is to the subcutaneous tissues of the back and this journey is accomplished fairly quickly. The exact route followed by the larvae in this latter migration varies, but some may find an intermediate resting place in the spinal canal. Once arrived at the back, the larvae enter upon the third stage which lasts about five

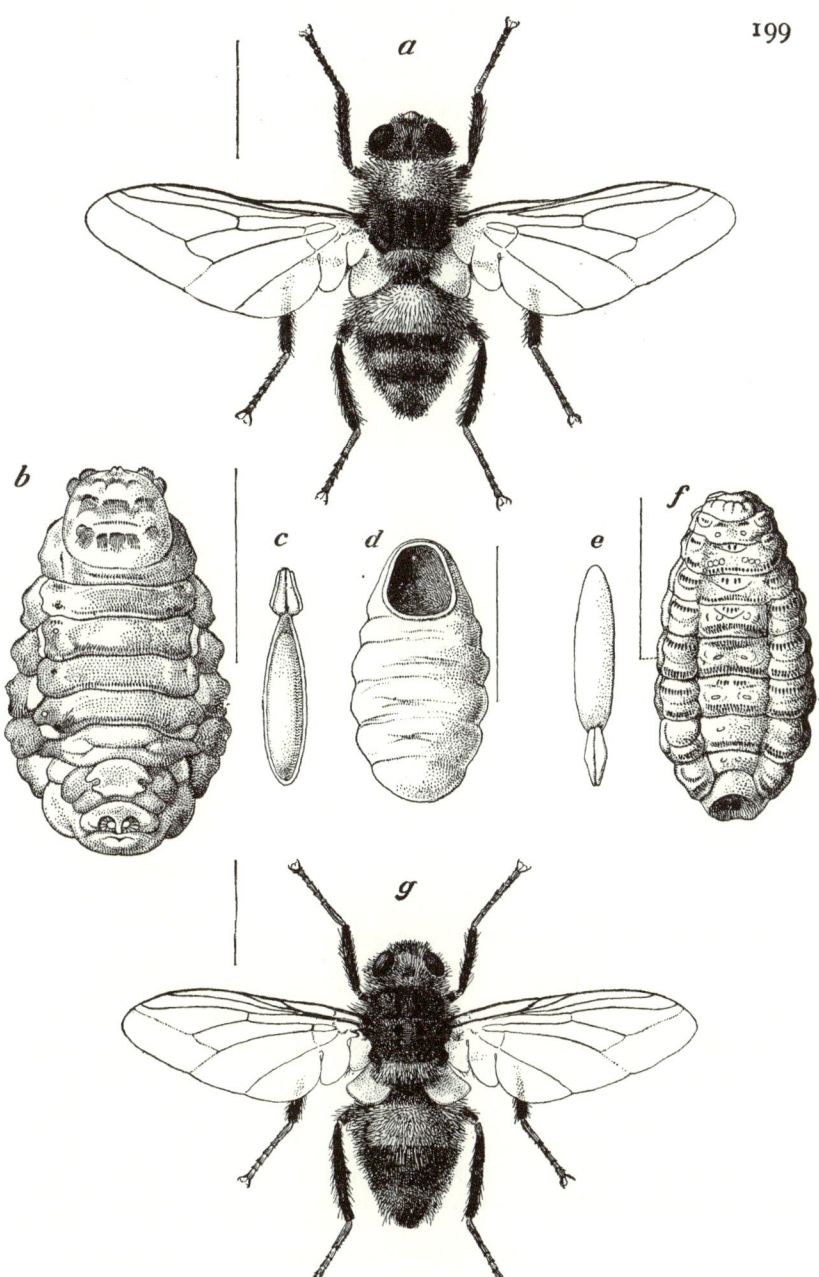

Fig. 53. Ox Warble Flies. *Hypoderma bovis* and *H. lineatum*. *a*, adult of *Hypoderma bovis*; *b*, larva of *H. bovis*; *c*, egg of *H. bovis*; *d*, puparium of *H. bovis*; *e*, egg of *H. lineatum*; *f*, larva of *H. lineatum*; *g*, adult of *Hypoderma lineatum*. (After Theobald.)

days. Soon after arrival in the back a hole is cut in the hide of the host. The fourth and fifth stages in the larval development are then passed in the 'warble' or swelling in the back of the animal. Bishopp finds that the total developmental period in the back varies between 35 and 57 days. When full fed the larvae emerge through the breathing hole cut in the skin and fall to the ground where they pupate, the pupal stage lasting about 35 days. In England the larvae begin to appear in the backs of the cattle about the middle

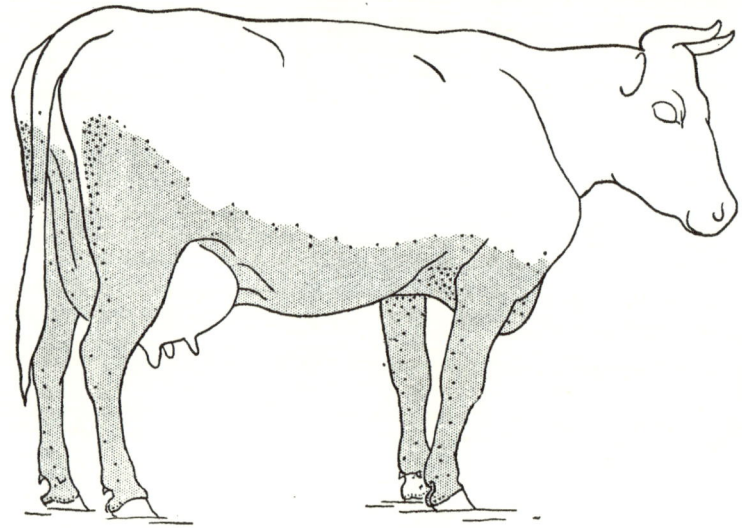

Fig. 54. This figure shows the regions on the animal where warble-fly eggs are deposited. The black spots indicate the places where most eggs are found, and the shaded parts the whole area where eggs may be deposited. The figure is for both species of warble flies. (After Hadwen.)

of February, or sometimes earlier, and are full grown and ready to leave the warble in April and May. The whole life cycle of the warble fly is thus seen to occupy about a year of which 9–11 months is spent as a larva within the body of the host.

Hypoderma bovis de G. Ox Warble Fly

DESCRIPTION. *Adult.* Male. A larger and stouter fly than *H. lineatum.* Head, frons broader than eye, eyes black, ocelli black, shining, orbits black; antennae sunk in cavity, yellowish in colour, third joint darker, arista short. Face with pale yellow pubescence. Thorax dull black, with four shining stripes which are partly obscured by yellow hair; scutellum black with close yellow hair. Abdomen, terminal hairs *yellow*. Wings, veins reddish brown to yellow; squamae white with brownish margins. Legs, femora and tibiae black, mainly black haired with some yellow; tarsi yellowish brown with black hairs.

Female. Similar, frons broader, abdomen broader with black band deeper. Length 13–14 mm. (Fig. 53, *a*.)

Egg. Similar to that of *H. lineatum* but rather larger; it is attached singly to the hair by a somewhat elbowed petiole. (Fig. 53, *c*.)

Larva. Very similar to the foregoing species, but is slightly larger and less spiny, the last *two* abdominal segments are without spines. The chief difference lies in the appearance of the stigmal plates. (Fig. 53, *b*.)

Puparium. Similar to that of *H. lineatum*. (Fig. 53, *d*.)

LIFE HISTORY. The life history of this species differs little from that of *H. lineatum*. Bishopp considers that the larvae of *H. bovis* seldom enter the gullet of the host. *H. bovis* seems only to oviposit in hot sunny weather, while *H. lineatum* will lay eggs freely at temperatures of 55° F. In England the adults of *H. bovis* appear 3–4 weeks later than the other species.

INJURIES TO HOST. The injuries caused to cattle by these flies are two-fold—firstly the loss of condition due to 'gadding' or the efforts of the animal to escape from the fly (*H. bovis*, especially, seems to arouse terror); and secondly the irritation caused by the larvae within the bodies of the host. Bishopp *et al*.[11] divide the second group into four sections as follows: (1) Soreness and pain produced by the penetration of the young larvae through the skin of the leg. (2) Irritation produced in the gullet and in other internal organs by the migrating larvae. (3) Inflammation produced along the spinal cord and on the main branches of the nervous system by the burrowing of the larvae along the spinal canal and at the ingress and egress of that canal. (4) The irritation produced by the later larval stages in the subdermal tissues of the back with accompanying pus formation.

DISTRIBUTION. Widely distributed in England and Wales, there appear to be no counties without the warble flies, though they are absent in the Scilly Isles. The counties of Hereford, Lincoln, Middlesex, Brecon and Radnor seem to be less affected; the flies are most prevalent in low-lying districts. Both species are widely distributed in Scotland and Ireland, and in the former country they are common in Midlothian, East Lothian, Berwickshire,

Forfarshire, Perthshire, Buteshire, Fifeshire, Morayshire, Banff-shire, and Aberdeenshire; while *H. lineatum* is recorded from Rox-burghshire and Selkirkshire and *H. bovis* from West Lothian, Kirkcudbrightshire, Wigtownshire, Ayrshire and Inverness-shire.

PREVENTIVE AND CONTROL MEASURES. Preventive measures lie in the provision of shelters in which the cattle may avoid attack by the fly, and in the application of chemicals to deter the insect from egg-laying. The second of these two measures has not as yet given satisfactory results, though driving cattle through a 2 per cent. solution of coal-tar creosote in water or processed crude petroleum is said to have prevented heavy infestation. (Imes and Schneider (18).) Control measures consist (*a*) in the destruction of the larva in the 'warble' in the backs of the cattle, which may be done either by extracting the ripe maggot by pressure or by means of forceps or by the application of larvicidal dressings through the breathing pore of the warble; (*b*) by the introduction of legislative measures (see p. 12). As regards the eradication of the larva in the warble, the application of dressings has been found the best and most practicable method. Various dressings are recommended by different workers. Bishopp *et al.*(11) have found the most effective ointments to be: derris, iodoform, pyrethrum, benzol and carbon tetrachloride in-jected into the grub cysts, or powdered tobacco and nicotine dust. As regards the last two substances, injurious effects are produced on the cattle if used at too great a strength. Sponging the affected area with a solution made by allowing 1½ lb. of fresh lime and 4 lb. of tobacco to stand in 1 gallon of water for 24 hours is recom-mended. After 2–3 days a fresh application should be made with freshly prepared fluid, followed by a third after a period of three weeks. Gaut and Walton (14) recommend the use of a derris soap wash, made up in the following proportions: powdered derris 1 lb., soft soap ¼ lb., water 1 gallon. Cattle should be examined and dressings applied four times during the season, the first at the end of March, the second in the third week of April, the third in the third week of May and the last dressing in the third week of June. In all these treatments systematic effort is necessary and the intervals between applications should not exceed 35 days. A good summary of all the available methods of control for warble fly will be found in the *Report of the Departmental Committee on Warble Fly Pest* (Ministry of Agriculture and Fisheries, 1926, H.M.

Stationery Office). The committee recommend the following dressings:

(1) *Tobacco powder and lime.* With this dressing it is important to use the right strength of nicotine and it is desirable that a tobacco with a standard nicotine content be used if possible.

(2) *Derris root.* This has an advantage over the above in that it is a powder and is more easily dealt with in carrying and mixing. Apply mixed with water either with or without soft soap.

(3) *Nicotine sulphate and lime.* Nicotine sulphate 2 fluid oz., calcium hydrate 1 lb., water 1 gallon.

Fluid dressings are best applied with a strong brass syringe and injected directly into the cyst. The ointment used in America, one part iodoform to five parts vaseline, is also recommended.

It cannot be emphasised too strongly that the co-operation of agriculturists is necessary in any measures to exterminate the warble fly; isolated effort is of little avail, while a few years' steady work on maggot destruction over any considerable area would do much towards the extermination of the pest.

As regards legislation against the warble fly, the objections to this, as stated on p. 12, are the lack of a reliable preventive and of any means of control over importation of warbles in cattle from other countries. The committee on warble control would, however, be prepared to advise the adoption of legislative measures, were the demand for them generally supported by agricultural opinion. Some success has been achieved by legislative measures in Denmark, Belgium and Switzerland, particularly in Denmark, where, however, the circumstances of agriculture differ very widely from those of the British Isles and where importation of cattle is a negligible factor.

Cephalomyia (Oestrus) ovis Linn. Sheep Nostril Fly

DESCRIPTION. *Adult.* Male. Head pale brown; frons narrower than eye; eyes brown; ocelli black, shining, very distinct; orbits yellowish with black dots; lower part of face with dense whitish pubescence; antennae with first two joints yellow, third joint black. Thorax brownish, densely covered with black wart-like tubercles, some long whitish hairs at humeri; scutellum brown, sparsely haired, tubercles large at apex. Abdomen black or brownish, marbled with white and a silvery lustre when alive, few hairs present on the dorsal surface, but some long white bristles on lateral margins. Wings clear and glassy, veins yellow, some black dots on veins at base of wing; extending beyond abdomen when closed. Halteres whitish, covered by the large alulae. Legs yellow with some black hairs.

Female. Frons broader than eye, tubercles fewer and larger than in the male. Length 10–11 mm. (Fig. 55.)

Egg. This insect is usually larviparous, but according to Theobald the egg is also sometimes deposited; he describes it as whitish and somewhat kidney-shaped.

Larva. The mature or third-stage larva is yellowish in colour, darkening posteriorly, somewhat convex dorsally and flattened ventrally. Dorsally on each segment except the first and last is

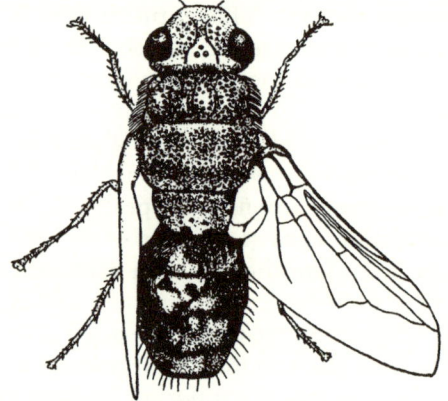

Fig. 55. *Cephalomyia ovis* Linn. The sheep nostril fly.
Much enlarged. (After Portchinsky.)

a transverse dark mark, on the under surface of the abdomen are numbers of backwardly directed spines which are smallest at the anterior end becoming progressively larger. The mouth-hooks are conspicuous, black or dark brown, curved. The caudal segment is truncated, sloping towards the anus, the posterior spiracles are fused into one stigmal plate bearing two sets of three parallel sinuous slits, below this the skin is tuberculate. There is a group of spines on a ridge situated on the anal segment. The posterior spiracles consist of two separate plates over which there is a dense chitinous network. Length 24 mm.

LIFE HISTORY. As stated above the fly is usually larviparous, depositing living larvae in the nostrils of sheep, generally during hot sunny weather. Theobald considers that eggs also are deposited occasionally in the same situation, though Portchinsky disputes this. The young larvae make their way through the nostrils into the frontal sinuses of the head where they attach themselves by hooks to the mucous membrane, remaining there for about nine months, the complete life cycle occupying ten months. In the second stage, according to Portchinsky (23), the larva loses all its bristles and hooks. This second stage lasts until the spring and early summer, when the third stage is entered upon. It is this third stage which gives rise to the fits in the affected sheep, known as 'false gid'. When full fed, the larvae release their hold and fall to the ground or are sneezed out by the animal. Pupation takes place in the soil often under stones or grass tufts, and the fly emerges after 21–60 days; it is short lived and does not feed. It is frequently to be found resting on walls and in shady places.

SYMPTOMS OF ATTACK AND INJURY TO HOST. The sheep attempts to evade the fly by keeping its head down, and to protect its nose by holding it close to the ground, generally in bare dusty places where the fly is more easily seen. This results in the nostrils becoming dusty and gives rise to catarrh. Affected sheep often exhibit a discharge round the nostrils, sometimes accompanied by sneezing, and they may have difficulty in breathing owing to the obstruction of the air passages.

TREATMENT. As a control measure, mercury bichloride at the rate of 1 part to 1000 parts of water injected into the nostrils is sometimes recommended, but preventive measures are the most practical ones. One method is to smear the sides of narrow salt troughs with tar so that the sheep is compelled to rub its nose in it to reach the salt. The tar is said to have a repellent action against the fly. Rapid rotation of pasture is a good practice, while Hadwen(15) recommends, in the case of small flocks, the provision of shelters such as "a dark shed with a curtain over the door", into which the sheep will go when the flies appear.

REFERENCES

DIPTERA (contd.)

Syrphidae

(1) COLE, F. R. (1929). Fumigation with calcium cyanide for the control of the greater and lesser bulb flies. *Journ. Econ. Entom.* XXII.

(2) COLLIN, J. E. (1920). *Eumerus strigatus* Fall. and *E. tuberculatus* Rond. *Entom. Mon. Mag.* LVI, No. 3.

(3) BROADBENT, B. M. (1927). Further observations on the life history, habits and control of the narcissus bulb fly. *Journ. Econ. Entom.* XX.

(4) FRYER, J. C. F. (1915). *The Daffodil Year Book.* Royal Hort. Soc.

(5) HODSON, W. E. H. (1926–27). The bionomics of the lesser bulb flies, *Eumerus strigatus* and *E. tuberculatus* Rnd., in S.W. England. *Bull. Entom. Res.* XVII.

(6) MacDOUGALL, R. S. (1913). Narcissus flies. *Journ. Board Agric.* XX.

(7) WILCOX, J. (1927). Observations on the life history, habits and control of the narcissus bulb fly. *19th Biennial Rept. Oregon Bd. Hort.* Portland, Ore.

(8) WILCOX, J. and MOTE, D. C. (1927). Observations on the life history, habits and control of the narcissus bulb fly in Oregon. *Journ. Econ. Entom.* XX.

Oestridae

(9) BEDFORD, G. A. H. (1925). The sheep nasal fly. *Journ. Dept. Agric. Union S. Africa,* XI, No. 2.

(10) BISHOPP, F. C. and DOVE, W. E. (1926). The horse bots and their control. *U.S. Dept. Agric. Farmers' Bull.* No. 1503.

(11) BISHOPP, F. C., LAAKE, E. W., BRUNDRETT, H. M. and WELLS, R. W. (1926). The cattle grubs, or ox warbles, their biologies and suggestions for control. *U.S. Dept. Agric. Bull.* No. 1369.

(12) CAMERON, A. E. (1922). Bot anaphylaxis. *Journ. Amer. Vet. Med. Assoc.*

(13) CARPENTER, G. H. and HEWITT, T. R. (1914). Some new observations on the life history of warble flies. *Irish Naturalist,* XXIII.

(14) GAUT, R. C. and WALTON, C. L. (1929). Ox warble fly. *Report on the demonstration and experiments carried out in Worcestershire,* Dept. Agric. Educ. Shirehall.

(14a) GAUT, R. C. (1930). Ox warble fly. *Ibid.*

(15) HADWEN, S. (1923). Insects affecting livestock. *Dom. of Canada, Dept. Agric. Entom. Br. Bull.* No. 29, N.S.

(16) HADWEN, S. and CAMERON, A. E. (1918). A contribution to the knowledge of bot flies. *Bull. Entom. Res.* IX, No. 2.

(17) HALL, M. C. and AVERY, L. (1919). *Journ. Amer. Vet. Med. Assoc.*

(18) IMES, M. and SCHNEIDER, F. H. (1921). *Journ. Amer. Vet. Med. Assoc.*

(19) LAAKE, E. W. (1921). Distinguishing characters of the larval stages of the ox warbles, *Hypoderma bovis* and *H. lineatum* with description of a new larval stage. *Journ. Agric. Res.* XXI, No. 7.

(20) LUNDBECK, W. (1927). *Diptera Danica,* VII.

(21) MacDOUGALL, R. S. (1919). The ox warble or ox bot flies. *Trans. of the High. and Agric. Soc. of Scotland,* XXXI.

(22) Ministry of Agriculture and Fisheries, London. Leaflet No. 118.

(23) PORTCHINSKY, J. A. (1913). *Oestrus ovis,* sa biologie et son rapport à l'homme. Mem. Bur. of Entom. and Sci. Committ. of Central Board of Land Admin. and Agric. St Petersburg, X, No. 3.

(24) THEOBALD, F. V. (1904). *Second report on economic zoology.* P. 132.

(25) WARBURTON, C. (1922). The warble flies of cattle, *Hypoderma bovis* and *H. lineatum. Parasitology,* XIV.

CHAPTER XIII

DIPTERA (*contd.*)

Family *Muscidae.* (*Calypterae.*)

Certain species of the 'Green-bottle' flies have acquired the habit of ovipositing on sheep in the region of open wounds or sores. The resulting maggots make their way into the animal where they feed on the flesh with disastrous results. In the British Isles *Lucilia sericata* Meig. is the chief culprit with possibly *L. caesar* Linn. A description of two 'Blue-bottle' flies *Calliphora erythrocephala* Meig. and *C. vomitoria* Linn. is also appended, though it is doubtful if these flies affect sheep in this country to a great extent.

Lucilia sericata Meig. Sheep Maggot Fly; Green-bottle Fly

DESCRIPTION. *Adult.* Male. Head, frons broad, orbits blackish above, rest of face silvery; antennae black, third joint about three times as long as the second; palpi yellowish. Thorax whitish, hoary, in front are three post-sutural acrostichal bristles. Abdomen, third segment without marginal bristles, forceps rather short. Wings yellowish, squamular piece at base of costa yellowish. Halteres yellowish. Legs black.

Female. Similar, frons broader than eye; thorax distinctly white and hoary in front with the beginnings of two faint median lines; abdomen bluish white, hoary, less shining. Length 6–10 mm. (Lundbeck.)

Egg. Yellowish white, elongate-oval, reticulate with cross ridges.

Larva. Usual muscid type, the posterior face of the caudal segment bears eight large and four small tubercles. Colour white to yellowish. Length 12–15 mm.

Lucilia caesar Linn. Green-bottle Fly

DESCRIPTION. *Adult.* Male. Head, antennae black; frons narrower than in *L. sericata*, third joint three times as long as second; palpi yellow. Thorax with two post-sutural acrostichal bristles. Abdomen, third segment without marginal bristles. Genitalia somewhat large, forceps rather long. Wings yellowish tinged, squamular piece at base of costa black. Halteres yellow. Legs black; femora bluish.

Female. Similar, frons nearly as broad as eye. Length 6–11 mm. (Lundbeck.)

Calliphora erythrocephala Meig. Blue-bottle Fly

DESCRIPTION. *Adult.* Male. Head, frons narrow above orbits, grey above becoming silvery; frontal stripe dark; lower part of face reddish, jowls with black bristles; antennae black, third joint long and black hairs below. Thorax dull bluish with a white tinge; there are four indeterminate dark stripes. Abdomen blue, rather shining, third and fourth segments with marginal rows of bristles. Wings hyaline darker at base; veins black or dark brown. Halteres dark. Legs black.

Female. Similar, frons broader than eye. Length 7–12 mm.

Calliphora vomitoria Linn. Blue-bottle Fly

DESCRIPTION. *Adult.* Very similar to the foregoing; jowls with black hairs above but red hairs below. Thorax, legs and wings similar to C. *erythrocephala*. Abdomen with marginal bristles on fourth segment smaller in this species. Length 9–12 mm.

LIFE HISTORY. The eggs are deposited on the wool of sheep, particularly near sores or open wounds. The eggs hatch very rapidly, in 24–48 hours. The maggots feed at first externally, later entering the flesh, the life cycle occupying 25–40 days according to the temperature. The flies are on the wing from May till autumn, the period of greatest infestation of the sheep being in August.

SYMPTOMS OF ATTACK. MacDougall (6) gives the chief symptoms as follows: matting of the wool fibres; discoloration of the wool where the maggots are at work; continual wagging of the tail; rubbing and biting by the sheep in efforts to allay irritation. There is usually much inflammation and oozing of matter from the sores together with a falling of wool.

CONTROL. The sheep should be kept as clean as possible, especially about the hind quarters, to avoid odours which help to attract the fly. Benzol has been found very suitable for killing the larvae in the wounds, while dressing the affected parts with a mixture of rape oil and turpentine, finishing with a dusting of sulphur, is also recommended. Parman and Laake (11) in America find that the best repellent substances to ward off the fly are certain pine derivatives such as pine-tar oil. Powdered pyrethrum, derris or copper carbonate are also good repellents. Carcases of dead animals and birds should be burned in order that they should not serve as breeding places for the fly. Holdaway (*Nature*, Vol. 126, No. 3182, 1930) has made a study of the biological agencies which play a part in regulating the numbers of sheep blowflies. See also Carew, J., *Queensland Agric. Journ.* XXXIII, pt 5, 1930.

NATURAL ENEMIES. Altson (1) gives a good account of two important parasites of blowflies; these are the Braconid *Alysia manducator* Panz. and the Chalcid *Nasonia brevicornis* Ashm., while the Ichneumonids *Phygadeuon speculator* Thoms. and *Atractodes bicolor* Gravenh. are recorded from *C. erythrocephala*. In France the Chalcid *Dibrachys affinis* Masi. has been bred from *C. erythrocephala* and *C. vomitoria* and the Ichneumonid *Theronia atalanta* Poda. from the latter. In Western Australia, the Chalcid *Stenoterys fulvoventralis* Dodd. is recorded (Newman and Andrewartha, *Journ. Dept. Agric. W. Aust.* (2), VII, No. 1, 1930).

Stomoxys calcitrans Linn. The Stable Fly; Biting House Fly

DESCRIPTION. *Adult.* Male. Head, frons narrower than eye, ocelli distinct with a tuft of divergent bristles; vertex with two parallel black marks, each with a row of inwardly directed bristles on its outer side; antennae dark brown or black, third joint rounded at tip; arista with a row of bristles; face with whitish pubescence. Palps short, proboscis very long black and shining, forwardly directed. Thorax greyish brown with four parallel black stripes, scutellum with a number of bristles. Abdomen greyish-brown with a broad dorsal stripe, two indeterminate spots on each segment, sutures black, venter pale. Wings brownish, hyaline, a few bristles present on the proximal end of the third longitudinal vein, apical cell widely open. Legs dark brown or black.

Female. Slightly larger, general colour lighter, frons broader than eye, abdomen with spots smaller and less defined; wings slightly grey, front tibiae brownish at base. Length 7 mm.

Egg. White, curved, rather bean-shaped, with a broad groove along the straighter side.

Larva. Very similar in general appearance to the larva of the common house fly, but differs from it in the arrangement of the stigmal plates which are smaller and wider apart. In *S. calcitrans* the chitinous periphery of the stigmal plate is roughly triangular in shape, with the adjacent sides straightened and parallel, instead of saucer-shaped, and the stigmal openings are S-shaped instead of sinuous. Length 11–12 mm. (MacGregor (7).)

LIFE HISTORY. In *Stomoxys*, as with all blood-sucking muscids, both sexes suck blood. The flies feed on practically any warm-blooded animals, such as horses, mules and cattle; they also attack man, dogs, cats and even chickens. The eggs are deposited in manure and stable refuse, particularly heaps of rotting straw and sometimes in grass mowings and sewage beds. Bishopp (2) gives the following list of breeding places in order of preference: straw of oats, barley, wheat, cow and horse manure. The eggs are deposited either singly or in batches up to 30 or more. The length of the egg-stage varies 1–4 days, according to temperature and humidity, the larval stage occupies 2–3 weeks, sometimes 7–8 days under optimum conditions; 9–13 days are spent as a pupa. The adults attack cattle usually on the outside of the fore leg, just below the knee, while on horses they feed on the lower parts of the legs and under the belly. *Stomoxys* shares with certain other blood-sucking muscids, such as the tsetse fly, the habit of gorging itself with blood until flight is no longer possible. Such gorged individuals are commonly found resting on farm buildings on hot summer days, exuding a drop of clear fluid from the anus at regular intervals.

DISTRIBUTION. This insect rivals the house fly in its universal distribution. It is probably native to the palaearctic regions, from whence it has followed man in his migration to all parts of the world. Not equally abundant wherever it occurs, it is more common in temperate regions. In the tropics it occurs very generally, but almost always in lesser numbers than in cooler climates.

CONTROL. The presence or absence of moisture in the breeding places has a great bearing on development, the larvae being able to withstand moisture almost up to saturation point. The young larvae, however, are more susceptible to extremes of moisture and temperature. Light also has a marked influence on the development of the larvae, particularly when small. Preventive and remedial measures consist in the removal so far as possible of manure heaps which act as breeding places, in the use of chemicals applied to the manure heaps to destroy the larvae and in deterrent sprays. A solution of potassium permanganate is sometimes recommended for use in stables, cattle yards and on dung heaps, also borax or colemanite (crude calcium borate) used as follows: 62 lb. borax and 0·75 lb. colemanite to 10 cubic feet or 8 bushels of manure; after application, 2–3 gallons of water should be sprinkled over the manure. As a repellent spray, Haseman (4) recommends 1⅔ gallons coal oil or crude oil and 1 lb. of flaked naphthalene, used as a very fine mist spray. It must be used sparingly and with care as it is liable to cause injury to animals. A less harmful spray would be one made of pyrethrum extract, which has given good results.

NATURAL ENEMIES. The Chalcid *Nasonia brevicornis* Ashm. has been bred from *S. calcitrans*.

The following list of Hymenopterous parasites have been recorded from Muscoid flies, including those dealt with here (Graham Smith(3)):

Cynipidae
  Diranchis sp.
  Kleidotoma sp.
  Figites sp.
Proctotrypidae
  Trichopria sp. possibly *elongata* Thoms.
  Aneurrhynchus sp.
  Conostigmus sp.
Ichneumonidae
  Phygadeuon speculator Thoms.
  Atractodes bicolor Gravenh.

Chalcididae
  Melittobia acasta Walk.
  Dibrachys cavus.
  Nasonia brevicornis Ashm.
  Muscidifurax raptor.
  Necremnus leucarthros Thoms.
  Spalangia hiota Hal.
Braconidae
  Alysia manducator Panz.
  Aphaereta cephalotes Hal.
  Aspilota nervosa Hal.

Family *Anthomyidae*

Arista plumose, pubescent or bare. Median wing nervure has a straight or evenly curved course, not bending abruptly upwards as in the Muscidae. Some members of this family are destructive pests of crops and are dealt with in detail.

Hylemyia coarctata Fall. Wheat Bulb Fly

DESCRIPTION. *Adult*. Male. Head, frons very narrow; orbits with short silvery pubescence; frontal stripe black; antennae and palpi black; arista with hairs of medium length; jowls smoky grey. Thorax ochreous grey with some indeterminate black markings; sides rather lighter. Abdomen rather narrow, darker than thorax, hairy with some obscure transverse striae and a faint interrupted dorsal stripe. Wings yellowish, with narrow yellowish brown veins. Halteres pale; costal spine small. Legs black; tibiae piceous.

Female. Frons as broad as, or broader than, eye; frontal space wide, reddish; thorax and abdomen pale, without markings; legs with four posterior femora and all the tibiae pale. According to Seguy(41) the second segment of the antennae has a spot of rusty red in front. Length 6–8 mm.

Egg. Ivory or creamy white in colour, slightly concave on the ventral, and convex on the dorsal side, pointed at one end, flattened at the other, with fine longitudinal surface markings in the form of minute ridges and curves which run longitudinally into one another. (Gemmill(23).) Length 1·8 mm.

Larva. (Fig. 56.) The newly hatched larva differs from the later stages by having *two* pairs of *unserrated* cephalic hooks, one pair above the other; the later larva has *one* pair of *serrated* cephalic hooks. There are probably three instars; the anterior spiracles possess 7–8 undivided papillae. The caudal end is truncated and bevelled dorso-ventrally, and has a characteristic arrangement of spines. (Fig. 58.) Length 20 mm.

Puparium. This is of the usual Anthomyid type and is illustrated in Fig. 57.

LIFE HISTORY. There is only one brood of *H. coarctata* in the year, the adults appear in June and occur until the end of September, the males usually appearing first. The life history of this fly is unusual in that the females oviposit in bare fallow or land sparsely covered with crops, such as roots or late potatoes; oviposition seldom occurs among cereals or in pasture. The flies prefer a light sandy soil to clay for egg-laying and will not oviposit

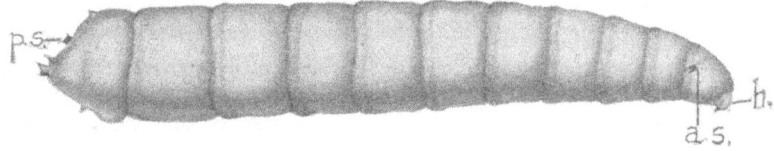

Fig. 56. Larva of *Hylemyia coarctata* Fall. The Wheat Bulb Fly. *h.* head; *a.s. p.s.* anterior and posterior spiracles. Much enlarged. (After Imms.)

in low-lying waterlogged land. Each female lays about 50 eggs, which are deposited singly in cracks in the soil. The winter is thus passed in the egg stage in the soil, the larvae emerging in late January and early February and seeking out the young wheat plants in which they feed until May. The time of hatching of the eggs is important since spring wheat sown towards the end of February is unlikely to be affected. Pupation takes place in the soil, the adults emerging in July, after about a month in the pupal state (Gough (1946), *Bull. Entom. Res.* XXXVII, 251–71).

CULTIVATED HOST PLANTS. Chiefly winter wheat, this fly has been known to attack winter rye and barley; oats are not attacked.

WILD HOST PLANTS. The chief wild host plant is couch grass (*Agropyron repens*).

INJURY TO HOST PLANT AND SYMPTOMS OF ATTACK. The damage to the wheat plant is caused by a single maggot feeding in the young shoot and devouring the growing point. The injured plants often occur in patches, which may increase in size through the migration of the larvae from plant to plant. An attack first becomes noticeable

about March or April, and can be recognized by the yellowing of
the central leaf, which later flags and dies. (Fig. 59.) Attack by
H. coarctata may be distinguished from wireworm attack (p. 106)
by the fact that in the latter case the outer leaves of the shoot as
well as the central one turns yellow.

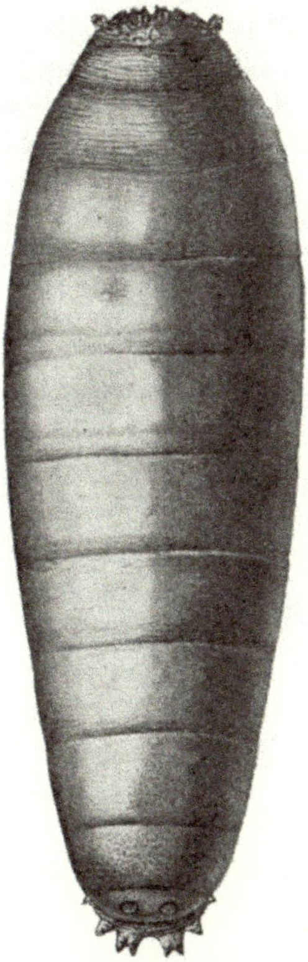

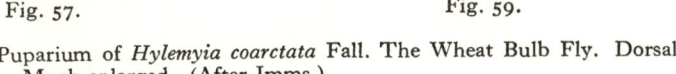

Fig. 58.

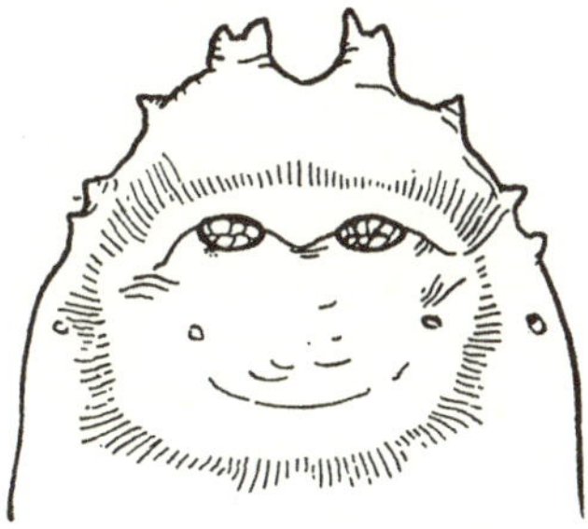

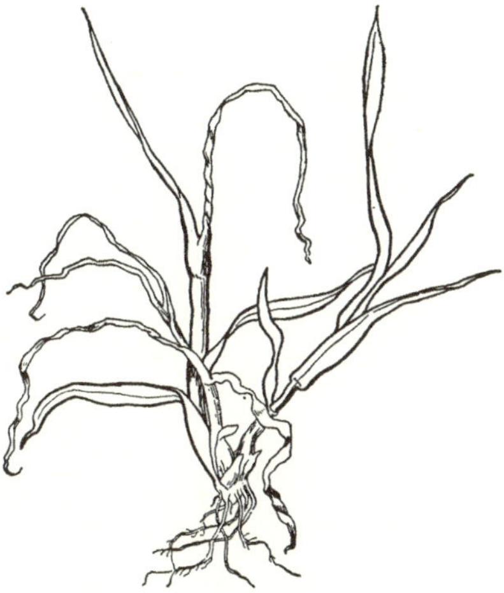

Fig. 57.   Fig. 59.

Fig. 57. Puparium of *Hylemyia coarctata* Fall. The Wheat Bulb Fly. Dorsal
view.   Much enlarged. (After Imms.)

Fig. 58. Larval caudal corona of *H. coarctata* Fall., showing arrangement of the
tubercles. Greatly enlarged. (After Rostrup.)

Fig. 59. Wheat plant damaged by the larva of *Hylemyia coarctata* Fall.; note
the destruction of the central leaf. (After Rostrup.)

DISTRIBUTION. Widely distributed in the British Isles, especially
in the Fen districts. The Ministry of Agriculture report severe
attacks by this insect in Cambridgeshire in 1927 and in Yorkshire
and the west of Berkshire in 1926.

CONTROL. The custom of drilling wheat after summer fallowing should not be followed in localities where *H. coarctata* is prevalent. In the Fens and other eastern districts, where damage occurs yearly from this fly, oats or barley are substituted for wheat. It should be realised, however, that barley, unlike oats, is also occasionally susceptible to attack. If possible, winter wheat or rye should follow grass, beans, some other cereal or mustard, before November if possible, and not potatoes or bare fallow. Early sowing of wheat to produce a strong plant before Christmas is recommended, also a liberal use of seed and the application of nitrate of soda to stimulate growth in the spring. Farmyard manure should be avoided if possible, as there is some evidence that it attracts the fly (Petherbridge, Stapley and Wood (1945), *Agric., Lond.*, LII, 351–4).

Hylemyia antiqua Meig. Onion Fly

DESCRIPTION. *Adult.* Male. Head, eyes contiguous, separated by a thin black line, orbits narrow and white; antennae of moderate

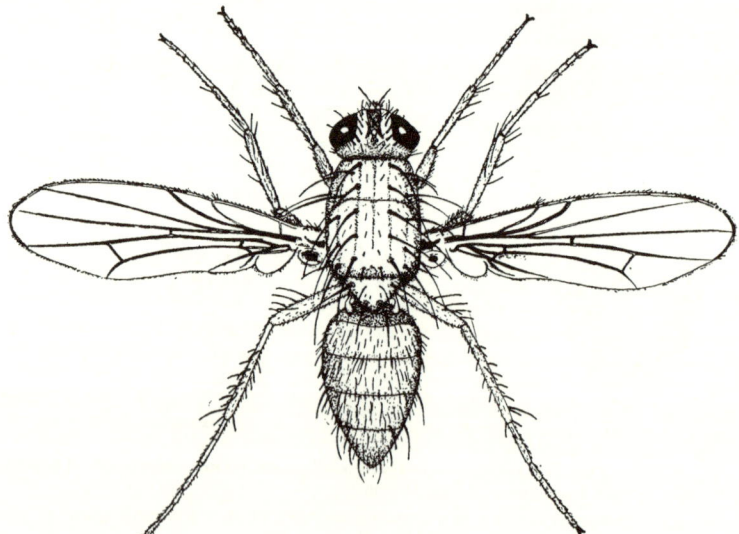

Fig. 60. *Hylemyia antiqua* Meig. The Onion Fly. Adult female. × 10.

length; arista thickened and pubescent at the base. Thorax unicolorous, yellowish grey with indistinct brown bands and four rows of black bristles. Abdomen oblong, greyish, clothed with black hairs, somewhat flattened and shorter than thorax, a thin median dorsal line is visible when seen from behind. Wings pale yellow; costal spine long. Halteres yellow. Legs piceous, third tibiae with a row of 6–8 short bristles of equal length, regularly spaced on the postero-internal face.

Female. Eyes widely separated, cheeks not reddish but some red in the intervening space between the eyes; abdominal tergites large; abdomen dull grey, more pointed than in the male. Length 5–7 mm. (Fig. 60.)

Egg. White in colour, chorion ridged, there is a shallow depression down one side extending about one third of its length. (Fig. 61.) Length 1 mm.

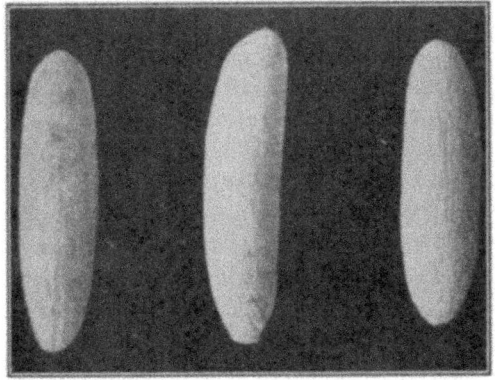

Fig. 61. Eggs of *Hylemyia antiqua* Meig. The Onion Fly. Much enlarged.

Larva. The larva is of the usual Anthomyid type, white and tapering. The anterior spiracles each contain as a rule eleven lobes but this number is not constant; there is a pair of small papillae situated anteriorly. The arrangement of the tubercles at the posterior end is important, being the only means of identifying the various Anthomyid larvae. This caudal corona in the onion fly larva consists firstly of eight dorsal, lateral and marginal tubercles, which do not seem to vary much in size or position in the various species of Anthomyid larvae; and secondly of two pairs of pointed tubercles with a pair of smaller ones in between. The larger of these last tubercles serve to distinguish the larva from that of the closely allied *H. brassicae*, in which the corresponding tubercles are bifid, and the pair of small tubercles are absent. Length 9–10 mm.

Puparium. Oval, dark brown, occasionally varying to lighter shades. The larval tubercles are easily discerned. Length 6–7 mm.

LIFE HISTORY. There are three broods of *H. antiqua* in a year, the third being often incomplete. There is no well-marked division of the generations, but each one overlaps the other, so that maggots in all stages, puparia and adults, are found throughout the summer. Adult flies, hatching from overwintering puparia, have been first noticed by the writer (44) on 8th May, though isolated specimens have been known to hatch during a mild spring much earlier than that date. The flies may be observed in large numbers ovipositing in the onion fields during the latter half of May. The eggs are deposited on the neck of the onion plant in clusters of half a dozen or more, and sometimes as many as 20–30 may be found together. They are deposited usually under the thin sheathing leaf surrounding the stem or in the crutch formed by the outside leaf and the stem. Occasionally eggs are deposited in cracks in the soil, but the more usual procedure is for the fly to lay them on the plant itself. The attachment is slight and eggs found on the surface of the soil have usually been detached by some external agency such as rain or wind. Maggots of the first generation have mostly pupated by the end of June. Flies of the second generation are on the wing in the second week of July, while third-generation adults were observed ovipositing in the third week of August. The larvae from these eggs pupate in the soil and thus pass the winter probably together with a certain proportion of late-hatched second-generation larvae.

CULTIVATED HOST PLANTS. Almost entirely restricted to the onion, the maggots occasionally attack leeks and shallots and have been recorded, possibly erroneously, as attacking tulip bulbs and lettuce.

WILD HOST PLANTS. No wild host of the maggot is known, but Baker (14) states that the flower of the dandelion (*Taraxacum officinale*) is an important food plant of the adult.

SYMPTOMS OF ATTACK AND INJURY TO HOST PLANT. The leaves of an attacked onion become soft and flaccid, later turning yellow and wilting, finally lying prone on the ground, the bulb in bad cases being reduced to a semi-liquid rotting mass. Any number of maggots from 3 or 4 up to 25–30 may occur in one onion bulb and where a large number of maggots are present the bulb may be entirely destroyed, only the portion above ground remaining. The most serious damage to the onion occurs in spring when the plant is still a seedling, the maggots eating through the stem just below the soil level.

DISTRIBUTION. Very common all over Great Britain, epidemics occur in certain years, especially in the midlands, where it is not uncommon for 90 per cent. of the crop to be destroyed.

CONTROL. Cultural methods of control consist in early sowing under glass and pricking out the young plants in April; small sets which have been stored over winter and then planted out in spring have shown themselves less liable to infestation. The writer has shown (45) that applications of nitrate of soda are beneficial in helping a crop to withstand attacks. Chemicals can only be used as deterrents; once oviposition has taken place and the maggots are in the bulb the application of chemicals is useless. Experiment has shown that chlor-cresylic acid is a good deterrent to egg-laying. The chemical should be applied in a 1 per cent. mixture with precipitated chalk or some similar inert carrier, the two should be well mixed together and dusted round the plants at fortnightly intervals, the first application to be given when the onions are about 1 in. above the ground. It is essential for the success of this method of control that the chlor-cresylic acid be applied before the eggs have been deposited. The 'poison bait' method of control is sometimes efficacious and consists in spraying the onions with a solution of sodium arsenite in water, with the addition of some treacle to attract the flies. Calomel applied to onion seed before sowing at the rate of 1 lb. per lb. of seed, effectively controls heavy onion fly infestations. Similar results can be obtained by two applications of 4 per cent. calomel dust applied along the rows after germination. Onion fly in *leeks* can also be controlled by treating the seed with calomel before sowing. Calomel-treated onion seed, sown directly after treatment, gives a better germination than untreated seed. Storage of treated seed should not be in a metal container (Wright (1939), *Agric.*, *Lond.*, XLVI, 147–54).

NATURAL ENEMIES. In Great Britain the chief parasites are the Braconid *Aphaereta cephalotes* Hal. and the Staphylinid beetle *Aleochara bilineata* Gyll. In Italy the Chalcids *Bothriothorax clavicornis* Thoms. and *Hygroplitis rugulosus* Thoms. and the Braconids *Microgaster globates* var. *anthomyarum* Bouché and *Rhogas rugulosus* Nees. are recorded. In Canada and America the

Cynipid *Cothonaspis gillettei* Wash. and the Braconid *Aphaereta muscae* Ashm. seem to be the most important parasites.

Hylemyia (*Chortophila*) *brassicae* Bché. Cabbage Root Fly

DESCRIPTION. *Adult. Male.* Head, eyes sub-contiguous, occupying the greater portion of the head; face somewhat prominent; arista pubescent. Thorax grey, dorsally there are three well-marked black bands (Meade prefers to call the thorax black with two short grey narrow stripes); scutellum black. Abdomen rather small, somewhat tapering with a clearly defined median black band and three transverse black lines, which divide the abdomen up into a number of small patches. Wings grey; nervures brown; costal spine distinct; third and fourth longitudinal veins converge slightly at their extremities. Legs black; third tarsi of the same length as the tibiae; there is a cluster of bristles on the hind femora.

Fig. 62. Eggs of *Hylemyia brassicae* Bché. The Cabbage Root Fly. Note the young larva emerging from the egg on the left, that on the right has already hatched as indicated by the ruptured shell. Much enlarged.

Female. Eyes widely separated; body colour lighter; abdomen broader and more pointed; legs brownish grey without the tuft of bristles present on the third femora of the male; as a whole less bristly. Length 5½–7 mm.

Egg. White, cylindrical, one end pointed the other somewhat truncated. Chorion sculptured into longitudinal ridges; down one side for three quarters of its length run two well-marked parallel sutures. Between these sutures is a somewhat depressed area in which the sculpturing is less pronounced than in the rest of the chorion. When the egg hatches, the smoother area between the sutures is pushed out, the larva emerging at the truncated end where the space between the sutures is widest. Length 1 mm. (Fig. 62.)

Larva. White, tapering to the anterior and truncated at the posterior end; integument smooth and shining. Anteriorly, the buccal armature has the lateral plate forked, while the anterior spiracles are fan-shaped with 11–13 processes. There are two pairs of tubercles surrounding the buccal opening. At the posterior end the caudal corona consists of six pairs of tubercles, while a seventh pair is situated ventrally near the anus. The outstanding feature of the caudal corona is the pair of bifid tubercles on the ventral rim of the last segment. In the centre of the caudal disc arise the knob-like posterior spiracles, which are brownish yellow in colour and heavily chitinised; they open by three slits. The newly hatched or first-stage larva differs from the later stages in having no anterior spiracles, while the posterior spiracles possess two openings instead of three. Length 8 mm. (Figs. 63, 64, 65.)

Fig. 63. Mature larva of *H. brassicae* Bché. The Cabbage Root Fly *in situ* in root of damaged radish. Much enlarged.

Puparium. Rather variable in size and colour, the latter ranging from deep red to light brown, the bifid tubercles of the larvae are still distinct. Length 6–7 mm. (Fig. 66.)

LIFE HISTORY. There are three generations of *H. brassicae* in the year, but there is considerable overlapping of the broods(46). The winter is passed in the pupal stage in the soil, the flies commencing to hatch about the beginning of May or a little later in northern

districts. Oviposition of the first generation continues throughout
May, the duration of the egg stage being 4–5 days. As a rule the
eggs are deposited on the stem of the plant, at or just below soil
level; the number deposited at one time varying from one or two
up to a dozen or more. On hatching, the young larvae pass down
through the soil and enter the root where they feed for an average
period of 23 days. When mature, the larvae leave the root and
pupate in the soil, the majority of the pupae being found within
a radius of 3–5 in. of the host plant. The
pupal stage of the generations other than
the hibernating brood varies in duration
from 15 to 35 days. Flies of the second
generation are on the wing at the end of
June, the period of maximum emergence
being roughly from 8th to 18th July.
The third-generation adults commence
to emerge about the end of July and continue to appear until the second week of
September. It is probable that a percentage of the second-generation pupae
do not hatch in the year of pupation but
remain until the following year.

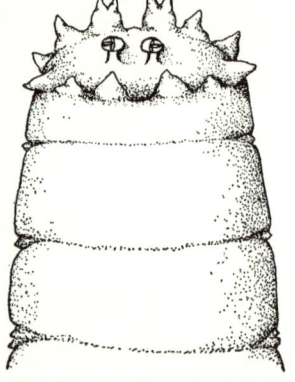

Fig. 64. Posterior segments of
mature larva of *H. brassicae*
Bché, the Cabbage Root
Fly, showing arrangement
of the tubercles; note the
two bifid tubercles which
assist in identification.
Much enlarged.

  CULTIVATED HOST PLANTS. The food
plants of *H. brassicae* seem to be confined
almost wholly to the Cruciferae. The
larvae occur commonly in cabbage, cauliflower, Brussels sprouts, kale, radish,
turnips and swedes, and there are records of occasional infestation
of stocks, beet and celery.

  WILD HOST PLANTS. Leaflet No. 122 of the Ministry of Agriculture gives several Cruciferous weeds as hosts of the maggot,
shepherd's purse (*Capsella bursa-pastoris*), charlock (*Sinapis arvensis*), and hedge mustard (*Sisymbrium officinale*), but the writer has
been unable to confirm this.

  SYMPTOMS OF ATTACK AND INJURY TO HOST PLANT. The symptoms
of maggot attack at the roots of Brassicas are characteristic; the
plant first of all becomes limp and flaccid, the leaves turning bluish,
later becoming yellow with complete collapse in a bad attack. On
hatching from the egg the young larva makes for the root just below

soil level, bores into the cortex and tunnels up and down. By throwing out adventitious rootlets, a plant is usually able to withstand attack by a few maggots only; but where a dozen or more are present, the whole root becomes hollowed out and the plant quickly dies. Although the root is the most usually attacked, maggots may also be found in the following places: (*a*) in the growing points of turnips and cauliflowers in which they cause 'blindness'; (*b*) in the inflorescence of cauliflowers; (*c*) in the midribs and larger leaf veins of Brassicas; this attack usually occurs in the autumn and is of little importance. It should not, however, be confused with a similar attack by the maggot of *P. rufipes* (p. 245), which normally tunnels in the midrib of Brassica leaves.

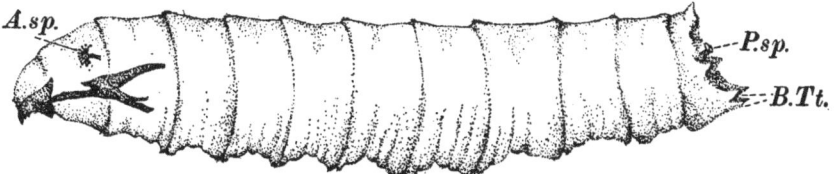

Fig. 65. Mature larva of *H. brassicae* Bché. The Cabbage Root Fly. *A. sp.* anterior spiracles; *P. sp.* posterior spiracles; *B.Tt.* bifid tubercles. Much enlarged.

DISTRIBUTION. Widely distributed throughout Great Britain. The Ministry of Agriculture report severe attacks in 1925 in the Bristol area, Cornwall and mid-Wales and again in the two latter localities the following year. In 1927 the fly was very abundant in Devon and Cornwall, south Yorkshire, mid-Wales and South Wales and the west midlands.

CONTROL. The writer(45) has had some success with the use of chemicals as deterrents, dusted along the rows before the fly has oviposited. Of these, two may be mentioned as having given fair control: (*a*) creosote, (*b*) chlor-cresylic acid, applied as a dust in a 1 per cent. mixture either with precipitated chalk or some similar inert carrier. In America and Canada much reliance is placed upon the use of corrosive sublimate (mercuric bichloride) used in solution (1 part of the sublimate to 1000 parts of water). The fluid is poured round the plant at the time of setting out and in the seed bed and thereafter at intervals of 2–3 weeks, about a cupful to each plant. The action of the sublimate, presumably, is to poison the larva during its passage from egg to root. Ingestion of the sublimate by the maggot is probably necessary for death to ensue, but the corrosive sublimate appears to have a toxic action on the egg and the very young maggot, possibly both a poisonous and contact effect.

Glasgow (25) has recently found that mercurous chloride, calomel, appears to have the same effect upon the larvae as does the mercuric salt. Four per cent. calomel dust with an inert filler is very toxic to insect eggs. The dust should be applied to the soil round the base of the plant. As much as possible should be deposited within 1 in. of the base of the stem, where the eggs are laid.

On the field scale, a duster, capable of giving definite 'puffs' of dust and not a continuous stream, should be used. The duster should be capable of holding several pounds of dust and be equipped with a long delivery lance to enable the worker to remain upright while using it.

In gardens and allotments the dust is best applied with a teaspoon, about half a teaspoonful, enough to form a complete collar round the base of the plant, at each application. The following recommendations are made:

(1) Brassica plants set out *before the last week of April* should receive an application of calomel dust in the last week of April. Plants set out *during or after the last week of April* should receive an application of dust within 4 days of setting out.

(2) Cauliflowers should receive a single application of calomel dust at the rate of 45 lb. per 4840 plants.

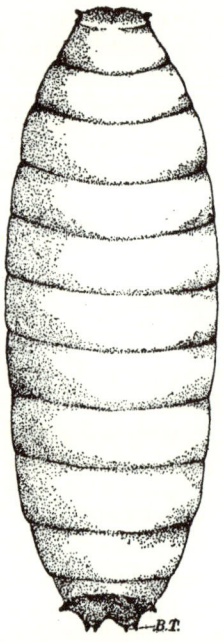

Fig. 66. Puparium of *H. brassicae* Bché, the Cabbage Root Fly. *B.T.* bifid tubercles of larval skin. Much enlarged.

(3) Brussels sprouts and broccoli should receive two applications of calomel dust, each at the rate of 30 lb. per 4840 plants. The second application should be put on two or three weeks after the first (Wright (1940), *Agric., Lond.*, XLVI, 765–72).

NATURAL ENEMIES. The following parasites and predators have been recorded from *H. brassicae* in Great Britain (46):

Coleoptera: *Aleochara bilineata* Gyll. The larvae of this Staphylinid beetle are parasitic upon the pupae. A closely allied species, *A. bipustulatus* (*nitida* Grav.), is probably also parasitic in the same manner.

Hymenoptera: The Cynipid, *Cothonaspis rapae* Westw., is the commonest and most important parasite of the cabbage root fly. The Braconid *Dacnusa stramineipes* Hal., and the Ichneumonids *Phygadeuon fumator* Grav., *Atractodes tenebricosus* Grav. (*vestalis* Hal.) are also recorded.

Diptera: The Anthomyid *Phaonia trimaculata* Bché. The larvae of this fly occur in company with, and feed upon, the larvae of *H. brassicae*. They may be distinguished from the latter by their larger size and the absence of the well-marked caudal tubercles.

In Europe the Cynipids *Cothonaspis gerasimovi* Meyer, *Figites anthomyarum* Bché, and the Braconid *Adelura dimidiata* Ashm. are recorded, and in America the Cynipid *Cothonaspis gillettei* Wash.

Hylemyia cilicrura Rond. (*Chortophila fusciceps* Zett.)

The larvae of *H. cilicrura*, the seed corn maggot of America, together with *H. striolata* Fl. and *H. trichodactyla* Rond., have been recorded in isolated cases as damaging the roots of runner and French beans in Kent, Devon and Lancashire. As the bean root fly is a serious pest of germinating maize, beans, etc., in America, it is wise to recognise the existence of such potential pests in the British Isles. In addition *H. cilicrura* attacks 'seed' potato tubers in America and the maggot has been shown by Leach [29] to be the vector of the bacterial disease of potatoes known as 'blackleg'. The larva carries the pathogene in the alimentary canal and the bacteria continue to exist throughout metamorphosis and are found still virulent in the intestine of the adult fly.

Phorbia genitalis Schnabl. (*Chortophila sepia* Meig.)

DESCRIPTION. *Adult.* Male. Head, frons very narrow; orbits narrow; cheeks with silvery white pubescence; antennae dark. Thorax dark grey with indeterminate longitudinal lines, acrostichal bristles well developed. Abdomen fuscous with a number of indistinct subquadrate markings on the dorsal surface, anal segments black and rather shining. Wings short with a yellowish tinge, base fuscous, a few hairs along the base of the costa. Legs dark brown to black with a row of spines on the upper surface of the hind tibiae.

Female. Frons wider than eye; frontal stripe black; orbits grey; abdomen dark grey. Length 3–5 mm.

Larva. The mature larva is white, the anterior spiracles are pale yellow and each consists of a wide rather flattened lobe, bearing about 22–25 papillae. The posterior spiracles are also pale yellow and project backwards almost horizontally; on the segment adjacent to them and situate dorsally is a pair of small conical tubercles or papillae, lateral to these and on each side is a pair of double papillae, while ventral to the posterior spiracles and equidistant from them is another pair of small conical papillae. Midway between, and slightly ventral to, the posterior spiracles is a small corrugated patch. The anal segment is slightly truncated ventrally and the anus opens between two rounded swellings; there is a marked absence of any well-defined caudal tubercles. On each segmental swelling, both dorsally and ventrally, is a patch of very small hooks or corrugations. Length 6–7 mm.

Puparium. Light brown in colour. No larval characters visible. Length 5 mm.

LIFE HISTORY. Little is known of the bionomics of *P. genitalis* in the British Isles and it does not appear to have been recognised as a pest of cereals in this country. In Siberia there is one generation a year, the adults appearing from overwintering puparia early in May. The larvae, which infest the young shoots of summer crops, appear about the 8th of June and are full grown about the middle of the month; they pupate in the soil close to the food plant. The eggs are deposited on wheat, particularly on the soft varieties; the presence of the larva in the wheat plant is indicated by a spiral tunnel running down the central leaf. Sakharov(37) finds that the harder varieties of wheat are more resistant than soft wheats, being protected by the ligula inside the leaf sheath. The food plant appears to be mainly wheat, though oats may be attacked as well. Late sowings of wheat appear to escape heavy infestation.

DISTRIBUTION. The larvae of this fly were reported as causing considerable damage in Cambridgeshire, Essex and Suffolk during the years 1922–4, and isolated attacks are reported from time to time in different parts of the country.

Pegomyia hyoscyami Panz. Beet or Mangold Fly

DESCRIPTION. *Adult.* Male. Head greyish; eyes medium size separated by a stripe; antennae short, somewhat incurved, basal joints yellowish or black, second joint sometimes reddish. Palps yellow with black tips. Thorax pale, pre-alar bristles shorter than the supra-alar ones. Abdomen reddish with a testaceous tint. (The colour of this insect is very variable, depending to a large extent on the host plant of the larva.) Somewhat swollen at the apex, with a faint dorsal line. Hypopygium globular and distinct. Wings pale yellow without costal spine. Halteres yellow. Legs, femora and tibiae yellow, tarsi black; in the first pair of legs the femora are sometimes brownish.

Female. Eyes rounder; frontal stripe reddish yellow; palps yellow at the tips; first and second antennal segments red; abdomen greyish. Length 5–6 mm.

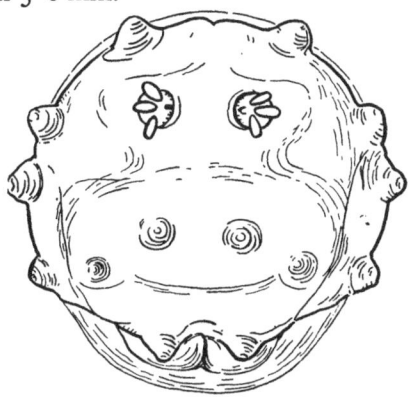

Fig. 67. Caudal corona of larva of *Pegomyia hyoscyami* Panz. The Mangold Fly. Showing arrangement of tubercles. Greatly enlarged. (After Kemner.)

Egg. White, elongate, not shining, truncate at one end, surface reticulate with hexagonal depressions, micropyle at truncate end as a small black spot. Length 0·85 mm.

Larva. Colour, dull yellowish or creamy. Mouth hooks sub-triangular with 4–6 teeth; anterior spiracles usually 8-lobed; caudal corona consists of seven pairs of tubercles. Cameron (17) divides the last segment into two areas, a dorsal-oblique and a ventral truncated area. The tubercles are arranged in a circle about the posterior spiracles, which have three apertures, the last pair of tubercles lying ventrally to the posterior spiracles. Length 7–9 mm. (Fig. 67.)

Puparium. Colour dark brown, at first paler; tapering towards the anterior end which is truncate. Larval tubercles not distinct, surface somewhat wrinkled. Length 4·5–5 mm.

LIFE HISTORY. There appear to be two biologic races of this fly, one of a paler colour which attacks henbane (*Hyoscyamus niger*) and is known as *P. hyoscyami*, and the other which attacks beet and

mangolds and is known as *P. hyoscyami* var. *betae* Curt. There does not
at present seem to be sufficient evidence to show that migration may
occur from weeds to cultivated crops, although Cameron (18) considers
that *P. hyoscyami* reared on belladonna can be induced to complete
its life history on mangolds. There are usually three overlapping
broods in the year, the winter being passed in the pupal stage, a
few inches below the surface of the soil. The fly emerges in April
and the female oviposits on the under surface of beet and mangold
leaves, the eggs being sometimes laid singly but more usually in
batches of 6–12, often deposited in parallel rows. The eggs hatch
in 3–5 days and the larvae feed in the soft parenchymatous tissue
of the leaves, in which they raise large blisters. When full grown
the larvae drop to the ground where they pupate, the pupal period
occupying 18–30 days. The whole life cycle occupies 36–40 days.

CULTIVATED HOST PLANTS. Cultivated crops attacked are mostly
mangolds, beet, sugar-beet, spinach.

WILD HOST PLANTS. The following weeds also serve as hosts for
the larvae, but, as already mentioned, it is not known how far
migration to cultivated crops may take place: henbane (*Hyoscyamus
niger*), deadly nightshade (*Atropa belladonna*), orache (*Atriplex* spp.),
white goosefoot (*Chenopodium album*), nettle-leaved goosefoot
(*Chenopodium murale*), Scotch thistle (*Onopordon acanthium*).

SYMPTOMS OF ATTACK AND INJURY TO HOST PLANT. The feeding
of the larvae on the parenchymatous tissue between the upper and
lower epidermis of the leaf produces a characteristic blistered
appearance. It is the first generation of larvae attacking the young
seedling plants which is of most importance, the leaf area being so
reduced that the young plants are unable to withstand the attack,
especially in hot weather.

DISTRIBUTION AND EFFECT OF CLIMATIC FACTORS. Widely distributed throughout the British Isles and Europe generally. The
attacks of this fly are usually intermittent, the species apparently
being subject to periods of depression and abundance. Some
recent German work (Blunck, Bremer and Kauffmann, 1929(16)) is of
interest in this connection. An analysis of the fluctuations of infestation in Germany indicates that the most important restricting
factors are summer temperature in its indirect effect on the development of parasites, and summer rainfall when it is definitely in
excess of the normal. Heavy rain washes away the eggs and kills

the pupae. In Central and Northern Europe, low summer temperatures prevent the coincident occurrence of parasites and host. *P. hyoscyami* is favoured by a cool summer when all the broods are equally numerous, in a hot summer the first brood is usually the most important. Epidemics are likely to occur when the temperature is such as to favour development of the fly and at the same time is cool enough to be unfavourable for an equally rapid development of the parasites. Local epidemics or outbreaks of *P. hyoscyami* depend rather upon climatic conditions in spring, which should not be too damp, followed by a low summer temperature. Local epidemics are often curtailed by heavy rain or a sudden rise in temperature.

CONTROL. Delayed drilling of seed, not before mid-May, helps to avoid the attack of the first generation. Rolling, if done before thinning, is fairly effective and does not harm the plants; deep ploughing in autumn tends to bring the puparia to the surface. As farmyard manure is said to attract *P. hyoscyami*, it should be applied in autumn to give it time to decay before the appearance of the fly. Schander and Gotze (39) recommend a poison bait to kill the adult flies before oviposition. This consists of 176 lb. of chaff mixed with 9–10 gallons of a solution containing 2·5 per cent. sugar and 4 per cent. sodium fluosilicate, these being the quantities per acre. A little horse dung is added to attract the flies and the bait should be placed in heaps of about 5 lb. each. (Roebuck, Baker and White (1945), *Ann. Appl. Biol.* XXXII, 164–70) recommend the following spray, sodium fluoride 4 lb., beet molasses 9 lb., water 10 gallons. The rate of spraying should be 25 gallons per acre.

NATURAL ENEMIES. The following parasites have been bred from *P. hyoscyami* in this country and on the continent:

Braconidae
Alysia divergens Bengt.
Apanteles congestus Nees.
Aspilota betae Bengt.
A. kemneri Bengt.
Diachasma bremeri Bengt.
Microgaster carinatus Bengt.
Opius betae Bengt.
O. carbonarius Nees.
O. fulvicollis Thoms.
O. nitidulator Nees.
O. procerus Wesm.
O. ruficeps Wesm.
O. spinaciae Thoms.

O. sylvaticus Hal.
O. wesmaeli Hal.
Phenocarpa pegomyiae Marsh.
Chalcididae
Decatoma betensis Dec.
Trichogramma evanescens Westw.
Ichneumonidae
Phygadeuon pegomyiae Haberm.
P. fumator Grav.
Diptera (Tachinidae)
Melanophora atra Macq.
Coleoptera
Aleochara bilineata Gyll.

Frost(22) records the following parasites in the United States of America:

Braconidae
 Alysia picta Goureau.
 Apanteles congestus Nees.
 Opius anthomyiae Ashm.
 O. foveolatus Ashm.
 O. pegomyiae Gahan.

Chalcididae
 Decatoma betae Decaux.
 Trichogramma minutus Riley.
Proctotrypidae
 Serphus sp.

Family *Psilidae*

Psila rosae F. Carrot Fly; Carrot Rust Fly

DESCRIPTION. *Adult. Male.* Head large, yellowish brown; antennae with the third joint half black, half yellow. On the vertex

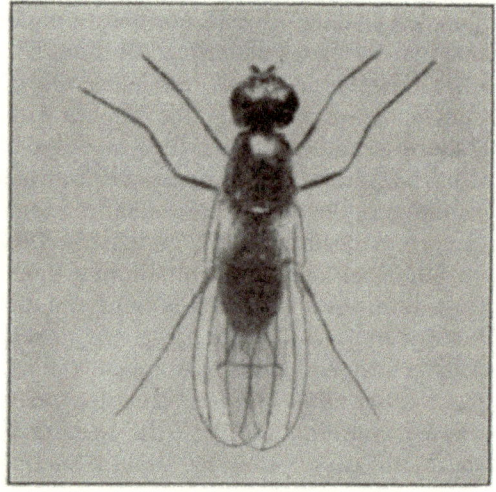

Fig. 68. *Psila rosae* F. The Carrot Fly. × 12.

is a black spot, slightly raised, bearing two bristles and surrounded by three ocelli; eyes black. Thorax black with two pairs of dorso-central bristles; these are important in identification. Abdomen greenish black, shining, metallic, sparsely covered with pale hairs. Wings iridescent with brownish yellow nervures; sub-costa absent. Legs pale yellow. (Fig. 68.)

Female. Similar, abdomen longer and tapering. Length 4–5 mm.

It should be noted that there are two closely similar species of fly, one at least of which may also attack the carrot, i.e. *Psila nigricornis* Meig. which differs from *P. rosae* in having all black antennae, and *Psila uniseta* (n.s.) which has the antennae rimmed with black and only one pair of dorso-central bristles (51).

Egg. Very small, white in colour with a delicate external sculpturing, micropyle distinct. Length 0·5 mm. (Fig. 69.)

Larva. The first-stage larva differs from the second and third stages in the possession of two large chitinous hooks which are lost in the second stage. The exact function of these hooks is not known but they are thought to assist the passage of the young larva through the soil. The third-stage larva is slender, shiny and semi-translucent, it is usually yellowish in colour due to the presence in the intestine of ingested carrot tissue. As a whole the body is smooth

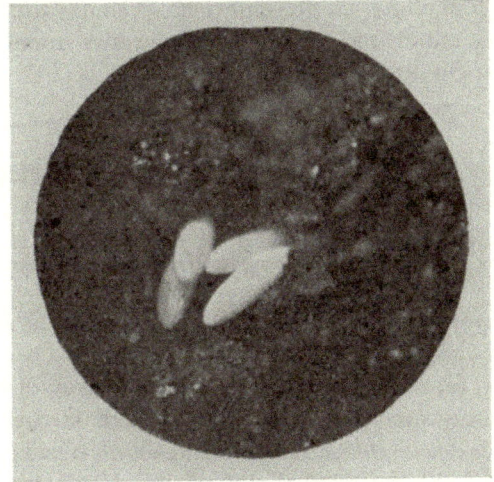

Fig. 69. Eggs of *P. rosae* F., The Carrot Fly, *in situ* in the soil surrounding the host plant. Greatly enlarged.

without projections, with the exception of the spiracles and some slight excrescences set with spines, one on the ventral surface of each segment. The posterior spiracles open by three lozenge-shaped apertures. Length 6–8 mm. (Fig. 70.)

Puparium. Cylindrical; pale yellow in colour, darkening towards the ends; the anterior end is much flattened, thickened and sloping. Length 5 mm.

LIFE HISTORY. There are usually two generations a year, the winter being passed in the pupal state in the soil. The flies commence to hatch from the overwintering puparia at the end of April and beginning of May, when they may be found on the flowers of cow parsnip (*Heracleum sphondylium* Linn.) and similar Umbelliferous weeds. Oviposition commences in May and reaches a maximum early in June. The eggs are deposited in cracks in the soil round the carrots (Fig. 69), and hatch in about 12 days. The young larvae usually make their way down to the lower end of the tap root and tunnel upwards, the majority of the larvae remain in the cortex of the carrot in a characteristic attitude, half buried in the root with their posterior segments protruding. The larvae are mature in about four weeks, after which they leave the carrot and pupate in the soil at a depth of about 2 in. The majority of puparia are to be found within a radius of 2 in. round the root. The pupal period is variable, larvae which pupate in July may continue to hatch all through the remainder of the summer. Usually the flies of the second brood commence to appear about two months after the appearance of the first.

Fig. 70. Mature larva of *P. rosae* F. The Carrot Fly. *A.* anus; *A.Sp.* anterior spiracles; *P.Sp.* posterior spiracles. Greatly enlarged.

CULTIVATED HOST PLANTS. In addition to carrots, *P. rosae* attacks celery, parsley, parsnips and occasionally turnips.

WILD HOST PLANTS. It is possible that certain Umbelliferous weeds which are attractive to the adults, such as cow parsnip or hogweed (*Heracleum sphondylium* Linn.), *Anthriscus* spp. and sweet cicely (*Myrrhis odorata* Scop.), may act as alternate hosts for the larvae, but confirmation of this is lacking, although the writer has experimentally induced the larvae to feed for short periods upon the roots of wild Umbelliferae.

SYMPTOMS OF ATTACK AND INJURY TO HOST PLANT. A carrot badly attacked by the larvae of *P. rosae* can be recognised by the wilting and yellowing of the tops. In cases of slight infestation, the plant remains sturdy, but the leaves become suffused with a coppery red tinge which is very characteristic. The damage done to the root is very extensive, the larvae boring mostly within the cortex; the presence of the larvae is indicated by the development of a rusty red colour on the carrot itself. The injuries to parsley, celery and parsnips are very similar, the damage being confined to the underground portions of the plants.

DISTRIBUTION. Very widely distributed all over the British Isles wherever carrots are grown. In Lancashire and Cheshire *P. rosae* is a perennial pest, it is scarcer along the sea coast.

CONTROL. The sodium fluoride-molasses mixture now in commercial use as a poison bait seems likely to be superseded in the near future by the modern insecticides. This bait is readily removed from carrot foliage by rain; on the other hand a D.D.T. spray of 0·5–1 per cent. concentration, having dried on the carrot foliage, should remain fully toxic to carrot flies alighting thereon for at least 3 weeks after application. The rapid growth of the carrot plants in spring would probably necessitate re-spraying after 14 days. Seven pounds of D.D.T. should be made into 140 gallons of emulsion, using solvent naphtha as the solvent. This emulsion, $\frac{1}{2}$ per cent. D.D.T. in strength, should be applied at the rate of about 50 gallons per acre. (Wright and Ashby (1945), *Bull. Entom. Res.* XXXVI, 253–68).

A recent report suggests (Newton, Satchell and Shaw (1946), *Nature, Lond.*, CLVIII, 417) that Gammexane may be more efficacious than D.D.T. in controlling carrot fly. Benzene hexachloride, containing 0·25 per cent. gamma isomer as crude hexachloride (formulated as P.P. flea beetle dust) gave control of a quite remarkable order. It was applied direct to the rows and heavy dressings did not appear to affect the carrot plants.

NATURAL ENEMIES. The writer has bred three Braconid parasites from the pupae of *P. rosae*; these are *Aphaereta cephalotes* Hal. and two undescribed species of *Dacnusa* (see *Insect Pests of the Horticulturist*, 1922, Benn Bros.).

Family *Trypetidae*

Lower fronto-orbital bristles present; oral vibrissae wanting; subcosta present, often indistinct; wings usually ornamented; ovipositor prominent, horny.

Acidia heraclei Linn. Celery Fly

DESCRIPTION. *Adult.* Male. General body colour variable, ranging from light brown to almost black. Head sub-globular; frons wide, brown; eyes deep green sometimes tinged with red, widely separated in both sexes; antennae pale yellow, third joint much larger than second; lower part of face pale; three ocelli triangularly placed on vertex; head bears a number of long bristles. Thorax black or dark brown, shining; the scutellum bears four long bristles and may be pale yellow or black, irrespective of the colour of the fly. Abdomen elliptical, rather shining, slightly broader than thorax. Wings broad, banded with brown wave-like markings,

with hyaline areas between the bands. Halteres yellowish. Legs
yellow; front tibiae with a row of spines on the inner face.

Female. Rather larger; abdomen broader and somewhat
pyriform; ovipositor conical and corneous. (Fig. 71.) Length
5 mm.

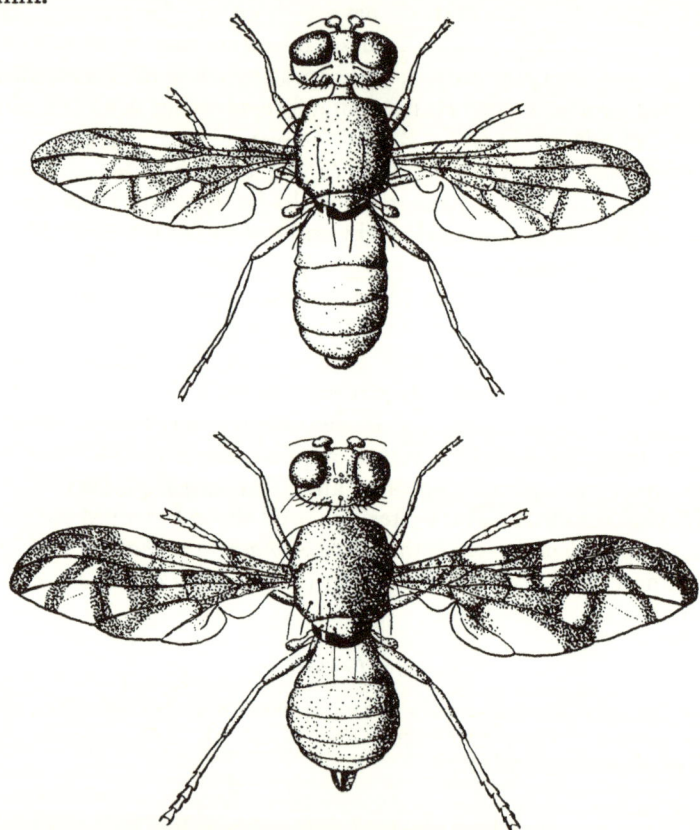

Fig. 71. *Acidia heraclei* Linn. The Celery Fly. Adult flies, male above,
female below. Greatly enlarged. (After Lundblad.)

Egg. White, elongate-oval, surface smooth, unsculptured;
micropyle distinct, mushroom-shaped. Length 0·5 mm.

Larva. White, somewhat glistening with a greenish tinge. On
the first segment above the mouth are two pairs of small sensory
organs. The eleven segments posterior to the cephalic segment
bear patches of locomotory spines. The anterior spiracles consist
usually of about 18 small lobes. Mandibular sclerite with three

DIPTERA 233

distinct hooks; posterior sclerite thin and widely forked. The posterior segment is truncated ventrally and there is a flat prominence which bears the posterior spiracles, each of which opens by three lozenge-shaped slits. (Fig. 72.) Length 7 mm.

Puparium. Light yellow, oval, slightly flattened dorso-ventrally, with deep dividing lines which give it a wrinkled appearance. Length 5 mm.

LIFE HISTORY. There are two generations in the year, the adults emerging in late April or early May from the overwintering puparia in the soil. The flies of the first brood are on the wing until the end of June. The eggs are deposited usually on the under surface of the leaf, but sometimes also on the upper surface, being inserted beneath the epidermis. The young larvae emerge in 6–10 days and commence to mine the leaf, they live gregariously, producing an irregular blotch type of mine which may occupy a

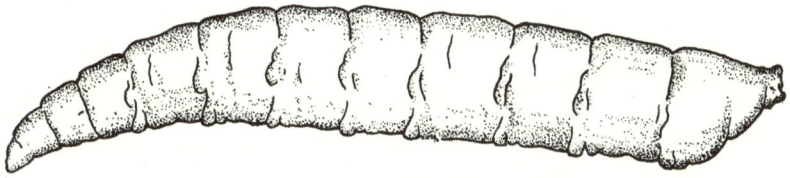

Fig. 72. Mature larva of *A. heraclei* Linn. The Celery Fly. Much enlarged.
(After Frost.)

considerable area. The larvae are mature in about three weeks and pupate either in the mine or in the soil. The summer pupal period lasts 25–30 days and the second brood of flies is on the wing from August till September, while the larvae of this generation occur from August till October and occasionally until November or December.

CULTIVATED HOST PLANTS. The chief cultivated host plant is the celery (*Apium graveolens*), but parsnips (*Pastinaca sativa*) are also attacked.

WILD HOST PLANTS. Frost(54) gives a list of wild host plants, chiefly Compositae and Umbelliferae; of the latter, various species of *Heracleum* and *Angelica* are the most important.

SYMPTOMS OF ATTACK AND INJURY TO HOST PLANT. An infested celery plant can be immediately recognised by the large blotchy type of blister which may be filled with moisture and contains the black excrement of the larvae. Later the attacked leaves become brown and curled, giving the plant a scorched appearance. In severe attacks, from 90 to 100 per cent. of the leaves in a bed of celery might be infested; in such a case the crop is of course entirely worthless.

DISTRIBUTION. Very widely distributed over the British Isles, it also occurs in France, Germany, Sweden and Denmark and is recorded from North America.

CONTROL. There are no very successful remedial measures for *Acidia heraclei*, and the control of this pest is a subject which urgently needs attention. Efforts should be made to keep down such weeds as *Heracleum sphondylium* which may act as reservoirs of infection. It is sometimes worth while to allow parsnips to grow on into a second year as trap plants and when the leaves of these are full of larvae the plants should be burned. Spraying the plants with a nicotine sulphate-soap solution has been tried with some success, while application of crude naphthalene at the rate of 2 oz. per square yard is sometimes recommended. On a small scale hand picking of infested leaves and collection of the adults by sweeping with a butterfly net on sunny days are useful measures.

NATURAL ENEMIES. The following Hymenopterous parasites have been recorded from *Acidia heraclei*:

In Great Britain:

Braconidae
  Adelura apii Curt.
  Aspilota fuscicornis Hal.
  Opius cingulatus Wesm.
  Sigalphus flavipalpus Wesm.
  Alysia apii Curt.

Chalcididae
  Halticoptera flavicornis Spin.
  Pachylarthrus smaragdinus.

Ichneumonidae
  Hemiteles crassicornis Grav.

In Sweden:

Braconidae
  Blacus exilis Nees.

Chalcididae
  Derostenus sp.
  Halticoptera smaragdina Curt.

In Italy the Braconid *Opius pallidipes* Wesm. and in America, in addition to some of the above, the Braconids *Alysia tipulae* Perris. and *Syntomosphyrum* sp. have been recorded.

Family *Oscinidae* (*Chloropidae*)

Arista bare; oral vibrissae rarely present; median and anal cells absent. Very small light coloured flies; common among herbage by roadsides, headlands, etc.

Chlorops taeniopus Meig. The Gout Fly;
Ribbon-Footed Corn Fly

DESCRIPTION. *Adult.* Male. Head yellow, occiput dark, a black triangular patch about the ocelli; frons brownish yellow; antennae black, yellow at base; lower part of face paler, eyes green and rather large. Thorax yellow, with three broad, dark chocolate-brown bands; scutellum much paler, whitish, arched on its upper side. Abdomen brown dorsally with dark bands on the segments. Wings greyish, veins brown, when folded they extend beyond the abdomen. Legs pale; tips of the tarsi and tips of front tibiae black; front tarsi black at base.

Female. Larger than the male and of a greenish tint. Length 5 mm.

Egg. Whitish, elongate-oval, surface hexagonally sculptured, flattened ventrally without sculpturing but with a ventral groove. Length 1 mm.

Larva. The full grown larva is cylindrical, tapering rapidly towards the anterior end and more gradually towards the posterior. According to Frew (67) the head is small, with a smooth surface, lacking the usual chitinous ridges. Two prominent maxillary palps are present, each consisting of a small rounded group of sense papillae surrounded by a dark brown chitinous ring. The antennae are laterally, instead of anteriorly, placed. The first two thoracic segments are covered with rows of chitinous spines, while the third thoracic segment has an anterior band of spines only. Eight of the nine abdominal segments bear a band of chitinous spines round the anterior border, while the ninth segment bears the anus and the posterior spiracles situated at the apices of two papillae. Length 6·3 mm.

Puparium. Yellowish brown in colour, somewhat flattened.

LIFE HISTORY. There are two generations in the year, a summer and a winter brood. The adults emerge about the end of May, and the eggs are deposited singly, usually on the upper side of the leaves of spring barley. Not more than one egg is laid on each shoot as a rule. The eggs hatch in about eight days and the young larva makes its way down into the centre of the shoot; it then eats its way down one side of the developing ear and on as far as the first internode, leaving a characteristic groove. When full fed, after two moults, the larva turns round within the shoot, so that the head is now directed upwards; it ascends a short distance and then pupates. The pupal stage occupies about 36 days, from July to August, and in the latter month the second brood of flies appears. These oviposit mostly upon couch grass, but also occasionally upon winter wheat or barley and upon self-sown wheat and barley. The winter is passed in the larval state, head downwards in

A

Fig. 73 *A*. Barley plant damaged by the larvae of *Chlorops taeniopus* Meig. The Gout Fly. *A*, a plant of winter barley showing the appearance of a shoot attacked by gout fly (centre), compared with normal shoots on either side. (Adapted from Frew.)

the grass shoots. About the end of February or early in March, the larvae reverse their position and pupate, the adults emerging towards the end of May.

CULTIVATED HOST PLANTS. Chiefly wheat and barley and occasionally rye.

WILD HOST PLANTS. The most important and probably the only wild host plant is couch grass (*Agropyron repens*).

INJURY TO HOST PLANT AND SYMPTOMS OF ATTACK. An ear of barley attacked by the larva of *C. taeniopus* is destroyed so far as any practical purpose is concerned. The larva, in its passage downwards, eats away part of the developing ear, which never escapes from the ensheathing leaves and in consequence cannot ripen. The symptoms of attack differ according to the time of year. In the case of the winter brood, the presence of the larva in both barley and couch grass is shown by the stunting of the attacked shoot, accompanied by an increase in the breadth of the leaves, which are unusually green. There appears to be no stem, while the central leaves are spirally curled. The distortion produced by the summer brood is characteristic and, typically, is a stem of two or three internodes ending in the swollen ear imprisoned in its sheathing leaves; it is from this symptom that the popular name of 'gout fly' is derived. (Fig. 73 *A*.) The effect of the larva in the plant is to cause an almost immediate cessation of growth. Frew has found that the actual place where the egg is deposited has a bearing on the infestation of the shoot, owing to

B   *C*

Fig. 73 *B* & *C*. Barley plants damaged by the larvae of *Chlorops taeniopus*. *B*, the typical form of summer attack on barley, showing the imprisoned ear; *C*, ear and ear-bearing internode from a typical example of barley attacked by the summer generation of *Chlorops taeniopus*. Below the ear the food groove, in which pupation sometimes takes place, is indicated. (Adapted from Frew.)

238 DIPTERA

the invariable habit of the larva in migrating downwards. If an egg
is laid on a leaf below the base of the ear, the young larva cannot
reach the ear; such are referred to as *non-critical* leaves. A larva
hatching from an egg deposited on a leaf, the base of which arises
from the shoot above the base of the ear, migrates downwards
and enters the ear; such leaves are termed *critical* leaves. Leaves
situate on shoots half-way up the ear are known as *half-critical*;
eggs are found deposited on all three types of leaves.

DISTRIBUTION. Widely distributed over the British Isles, attacks
were recorded in Yorkshire and the west midland provinces in
1925, and in 1927 there were some serious attacks in Wiltshire.

CONTROL. As with most insect pests of agricultural crops,
remedial measures are of little avail; preventive measures, however, can be helpful in reducing infestation by the gout fly. Frew (68)
recommends suitable manuring, particularly superphosphates, as
having a marked beneficial effect. This is due to the stimulating
effect upon the maturing of the ear and the growth of the ear-
bearing internode. Small dressings of nitrogenous manures may
reduce infestation, but large dressings are unsuitable owing to a
tendency to retard the growth of the ear. Early sowing is advantageous, and this should be carried out at the earliest possible
date allowed by the soil conditions.

NATURAL ENEMIES. The Braconid *Coelenius niger* Nees. and the
Chalcid *Stenomalus micans* Ol. are endoparasites of the larvae;
while Frew records two unidentified Hymenoptera which are ecto-
parasites. Curtis gives the Ichneumonid *Sigalphus caudatus* Nees. as
parasitising *C. taeniopus* and Leonardi records the Braconid *Blacus
brachialis* Rond. in Italy. In addition the Braconid *Dacnusa areolaris*
Nees. and the Chalcid *Stenomalus lactus* are recorded from Europe.

Oscinella (Oscinis) frit Linn. Frit Fly; Oat Fly

DESCRIPTION. *Adult.* The following description of the adult flies
is quoted, slightly shortened, from Aldrich (59): Head, thorax and
abdomen black; frons usually wider than eye; frontal triangle
shining black, reaching usually almost to the root of the antennae,
but shorter in many specimens. Face black, rather concave, its
lower edge sometimes yellowish. Vertex bearing one outwardly
directed minute bristle near the corner of the eye, a minute con-
vergent pair of ocellar bristles and behind them a pair of also con-
vergent ones on the occiput. There is a row of three or four small

hairs along the inner edge of the eye in the opaque part of the front and a pair of hairs at the edge of the mouth as well; antennae black, third joint rather large and round; the arista with very minute pubescence beyond its basal fourth, often lighter in colour, rarely almost white; proboscis and palpi black. Thorax, dorsum sub-shiny to shiny with minute dark hairs usually arranged in rows lengthwise; one pair of minute dorso-central bristles before the scutellum, far apart; scutellum of ordinary form; pleurae shining, black without bristles. Halteres yellow. Abdomen black, sub-shiny, rarely the first segment yellowish. The black colour extends as far as the ventral surface, which is often paler. Legs, coxae and femora black; trochanters often yellowish; tibiae rarely entirely black, usually paler at base and tip, fore and middle tibiae often wholly yellow, hind ones with at least a black ring; tarsi yellow, darkened towards tips. Wings sub-hyaline, sometimes brownish, varying slightly in width, costa extending to fourth vein which ends slightly behind apex, anal angle well developed. In the male the genitalia often protrude, showing a pair of distinct claspers curved backwards, these may be retracted or invisible.

In the female the abdomen is pointed, ending in a minute pair of palp-like organs at the tip of the telescopic 3-jointed ovipositor. Length 1·1–2 mm. (Fig. 74 A.)

Egg. White, one side slightly more curved than the other, surface sculptured into a number of fine ridges running lengthwise. Length 0·7 mm.

Larva. Somewhat narrow in proportion to its length, tapering slightly towards the head end, where are the 1-jointed antennae. Situated below these are the sense organs consisting of a crescent of small prominences surrounding a central area which bears a spine. Dorsally the head region carries a number of spines directed backwards. The general body surface is smooth, the mouth hooks are short and strong with two teeth posterior to the apex. The anterior spiracles are 4-lobed while the posterior spiracles, which have three openings, are situated on prominent outgrowths of the segment directed backwards. Adjacent to the posterior spiracles are four sets of radiating thickened ridges which are probably protective to the spiracles. The anal segment projects ventrally from the spiracular segment and is very small and highly chitinised, with a semicircular anal proleg on each side. Length 3 mm. (Hewitt(72).)

Puparium. Yellowish at first, later turning to a red brown, the posterior spiracles and larval mouth hooks are clearly visible. Length 2·5 mm. (Fig. 74 B.)

LIFE HISTORY. There are probably three main generations in the year in Great Britain, the winter being spent as resting larvae in

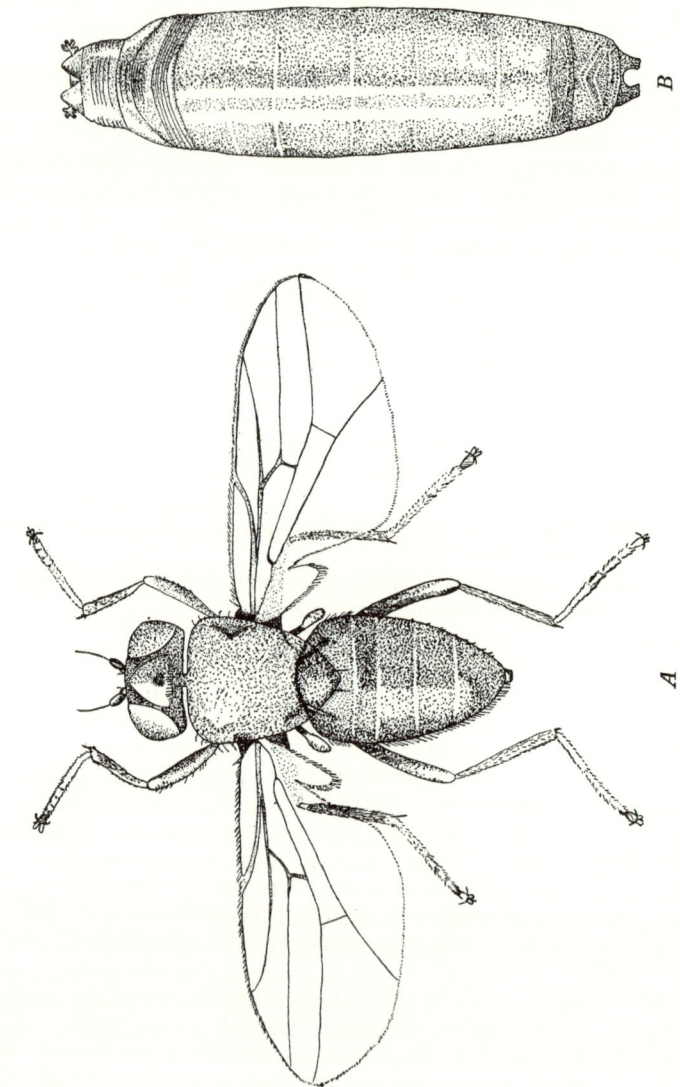

Fig. 74. *Oscinella frit* Linn. The Frit Fly. *A*, adult; *B*, puparium. Greatly enlarged. (After Aldrich.)

wild grasses, self-sown cereals or winter wheat. Pupation takes place in early spring, the adult flies commencing to emerge at the end of April or beginning of May, with a maximum emergence at the end of May. These flies oviposit on spring-sown oats, the eggs being usually deposited on the lower blades seldom more than three at a time. The eggs hatch in about four days and the young larvae make their way through the tissues in a horizontal spiral to the centre of the stem, where the supply of sap is greatest. This procedure destroys the central shoot, the death of which is a characteristic of frit fly attack. After feeding for about 15 days the larva pupates within the shoot, the whole life cycle of the spring brood occupying about 35 days. Flies of the second or summer brood commence to hatch at the end of May or beginning of June and continue hatching until the middle of August, with a maximum emergence about the middle of July. The eggs of this brood of flies are deposited within the glumes, and the resulting larvae feed upon the developing grain. The third brood of flies are on the wing from the end of July to the beginning of September or sometimes later, with a maximum emergence about the middle of August. This brood deposits its eggs mostly in wild grasses such as couch grass (*Agropyron repens*) and also upon volunteer or late-sown cereals, while it is possible that migration of larvae may take place from the wild grasses to winter wheat. The larvae of this brood feed for a while in the shoot, and then hibernate as resting larvae till the following spring, when they pupate.

CULTIVATED HOST PLANTS. In Great Britain oats are most commonly attacked, but infestation of barley, winter wheat, rye and maize also occurs. In Central Europe the fly seems to prefer barley, while Aldrich (59) in North America describes it as a pest of wheat, and very rarely on oats.

WILD HOST PLANTS. There are considerable numbers of wild perennial grasses which may act as alternate hosts for the larvae of the frit fly. Among these, the following are the more important: couch grass (*Agropyron repens*), *Setaria* sp., various rye grasses (*Lolium* sp.), fescue grass (*Festuca* sp.), tall oat grass (*Arrhenatherum avenaceum*), meadow grass (*Poa* sp.), and the common holcus (*Holcus lanatus* Linn.).

INJURY TO HOST PLANT AND SYMPTOMS OF ATTACK. The injury to a young oat plant by the larvae of the spring brood consists mainly in the destruction of the central shoot which first of all turns pale green, then greenish yellow, and finally withers. (Fig. 75.) Cunliffe (61) considers that the degree of destruction of the central leaf, which is the characteristic external sign of attack, depends partly on the method of entry and subsequent working of the larva and partly on the state of development of the plant. In the case of the spring

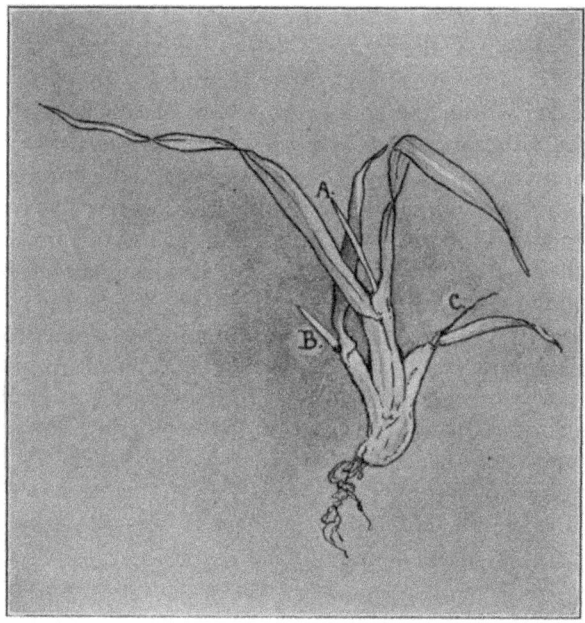

Fig. 75. Oat plant attacked by the larva of *O. frit* Linn. The Frit Fly. *A*, healthy tiller; *B*, attacked tiller, early stage; *C*, attacked tiller, late stage. (Reproduced by permission of the Ministry of Agriculture.)

attack, the plant is often induced to put out tillers which may in their turn become infested, with the result that a short stunted plant, with a number of shoots, is formed. The larvae of the second brood feed in the developing ear and destroy the grain. According to Fryer and Collin (69), the normal period of susceptibility in the oat grain begins very shortly before fertilisation and closes before the grain reaches half its ultimate size, with the probability that maximum susceptibility occurs at the time of fertilisation.

DISTRIBUTION. Very widely distributed throughout the British Isles, especially in the south of England from Cornwall to Kent, and the eastern districts except the Fens. It is an important cereal pest in north-western Europe and North America. Although very common all over the country, there occurs considerable fluctuation in the numbers of this fly. Fryer (*Insect Pests of Crops*, 1925–7) considers that the records of the past ten years suggest that the variation in losses in spring oats due to frit fly are of two different types. There is firstly a fluctuation not obviously connected with a change in the numbers of the frit flies and he suggests that this may be due to weather conditions acting on oat plant and perennial grasses and the fly in such a way as to transfer the attack from oats to grasses. The second variation in damage appears to be due to an actual reduction in the numbers of the fly, which may have been caused by parasitism, or to the interplay of some other controlling factor which results in rather long periods of abundance and the reverse.

CONTROL. Cultural methods of control, alone, are applicable to frit fly. Early sowing of spring oats and other cereals liable to attack, and late sowing of winter cereals are recommended. The work of Cunliffe, Fryer and others has shown that the oat plant is most susceptible in the 2-leaf and 3-leaf stages, once the plant has reached the 4-leaf stage it is less liable to attack. Oat plants, therefore, which have reached this latter stage in the spring before the appearance of the fly, are less liable to damage than younger plants. Cunliffe considers that, on this assumption, a solution of the problem would be obtained if the plant breeder could evolve a spring oat in which the rate of growth was so adjusted that coincidences between the most susceptible stages and the periods of maximum emergence of the fly did not occur. The most important contribution to the control of the frit fly would be the production by the plant breeder of an oat, immune or highly resistant to frit, and this is being attempted at the present time. The application of manures to stimulate plant growth is sometimes useful. Cunliffe [64] found that excess of nitrogenous manure failed to influence the extent of shoot infestation, but effected a reduction in the extent of grain infestation during a season when the extent of the attack of the fly was below normal. It is worth while testing different varieties of oats in an infested area to discover which variety is most resistant in that particular area. Winter cereals should not be sown after a grass ley, especially ryegrass (*Lolium perenne*) unless the field has been ploughed before harvest time; if it is ploughed later the cereals may be severely affected by the larvae that migrate from the buried grass. Winter oats suffer little from the larvae of the spring generation, but spring oats may be severely damaged. They are very resistant, however, after they have developed more than four leaves, and early sowing is the best way of enabling them to reach this stage before the adults of the overwintered generation emerge (Petherbridge and Weston (1945), *Agric., Lond.*, LII, 463–4). NATURAL ENEMIES. Cunliffe(60) records four species, while Imms(73) has bred six species, of parasitic Hymenoptera from *Oscinella frit* in England. Three of these latter, two Chalcids and a Braconid, were present very rarely, the remaining three are as follows:

Chalcididae
  Halticoptera fuscicornis Walk. (*Dicyclus fuscicornis* Walk.).

Proctotrypidae
  Loxotropa tritoma Thoms.
Cynipidae
  Rhoptromeris eucera Hartig.

In addition the following also occur in England:

Braconidae
  Sigalphus caudatus Nees.
  Aphidius granarius Marsh.

Chasmodon apterus Nees. (*Alysia aptera* Nees.).

The following are recorded in other countries:

In Germany:

Cynipidae
  Cothonaspis hexatoma Först. (*Hexaplasta fuscipes* Meyer).

In Russia:

Chalcididae
  Semiotellus nigripes Lind.
  Stenomalus micans Ol. (*Pteromalus micans* Ol.).
  Pteromalus puparum Linn.

In Poland:

Chalcididae
  Neochrysocharis immaculatus Kurd.

Chalcididae
  Halticoptera petiolata Thoms.
  Gonatocerus sulphuripes Först.
    (*Rachistus sulphuripes* Först).

  Trichomalus cristatus Först. (Recorded from Germany also.)
  Merisus intermedius Lind.
  Polycystus oscinidis Kurd.
Braconidae
  Gyrocampa pospelovi Kurd.

According to Imms the parasites of *O. frit* in England become more abundant as the season advances, with the result that the frit fly affecting late-sown oats suffers markedly heavier parasitisation than when it attacks oats drilled early in the season. Goodey(71) has recently described a Nematode worm, *Tylenchinema oscinellae* gen. et sp. n., which is parasitic in the adult fly. The effect of this worm, which is widely distributed in England and Wales, is to sterilise the host.

Family *Agromyzidae*

Phytomyza rufipes Meig.

DESCRIPTION. *Adult.* Male. Head brownish with some white markings round the eyes, bristles sparse; there is a small black raised area containing spines situate between the eyes; antennae very short and broad, yellow; arista long and bare. Thorax greyish with a number of long bristles sparsely set. Four pairs of long spines form two ill-defined rows down the centre of the dorsum; scutellum large. Halteres yellowish white. Abdomen greyish, thickly set with bristles. Wings with brown veins. Legs brown.

Female. Antennae black except for the extreme edge which is yellow. Length 4 mm.

Larva. Smooth, shining, with complete absence of tubercles; each segment in the mature larva is strongly differentiated by concentric rings of small spines surmounting each segmental

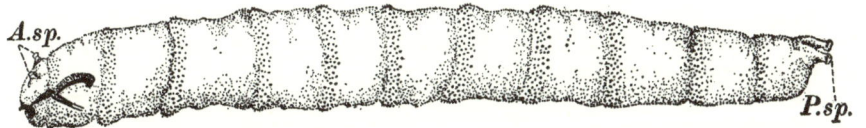

Fig. 76. Mature larva of *Phytomyza rufipes* Meig. *A.sp.* anterior spiracles; *P.sp.* posterior spiracles. Much enlarged.

swelling. The anterior spiracles stand well out from the body, and the posterior spiracles are directed downwards and backwards, forming a distinctive character by which the larva may be identified. Length 6–7 mm. (Fig. 76.)

Puparium. Light brown in colour, slender; most of the larval characteristics are still visible especially the spiracles. Length 4 mm.

LIFE HISTORY. There appear to be three or four generations in the season, adult flies being plentiful from May to October. The eggs are inserted in the tissue of the leaf blade of Brassicas, the place of oviposition being marked by a small whitish ring. The larvae live mostly in the midrib and leaf bases, occasionally being found in the stem and growing points of young plants. The winter is passed in the pupal state in the soil.

HOST PLANTS. Mostly Crucifers, especially cauliflowers and cabbages. The writer has not observed this insect on wild hosts.

SYMPTOMS OF ATTACK AND INJURY TO HOST PLANTS. The blackened borings of the larvae in the midribs of cauliflower leaves are plainly visible from the exterior. More serious damage is occasionally done to young cauliflowers through the larvae boring in the head and causing 'blindness'.

NATURAL ENEMIES. The larva is heavily parasitised by the Braconid *Dacnusa stramineipes* Halid.

Family *Hippoboscidae*

Head sunk in a depression in the thorax, antennae apparently 1-jointed and inserted in depressions; palps form a sheath round the proboscis, claws strong and sometimes toothed. Dorso-ventrally flattened with a leathery integument, wings either present or absent, ectoparasitic. Two species are shortly dealt with.

Hippobosca equina Linn. Forest Fly; Cattle Fly

DESCRIPTION. *Adult.* The whole insect is flattened, leathery and shining. Head, frons wide, black, orbits paler, face yellow, eyes and antennae black. Thorax much broader than head; anterior lateral margin yellowish brown, a patch of bristles at the humeri, dorsum shining black with a white border, in the centre a large triangular white spot with two smaller ones below, the third spot being on the scutellum, the three form an interrupted white line. Abdomen dark with pale hairs. Wings fully developed, dingy hyaline; front veins thick and prominent, brown. Legs yellowish brown, femora and posterior tibiae with black bands; claws black. Length 6–7 mm.

LIFE HISTORY. The adults appear in May and occur until August, they fly only in sunny weather. The larvae are nourished inside the body of the female and are deposited only when ready to pupate. According to Roberts (83) the larva is globular in shape and creamy white in colour at the time of extrusion, with a black cap and two conical projections at that end. It appears incapable of individual movement, and is without visible segmentation. Pupation takes place after a few hours from the time of deposition, the larval integument becoming thickened and dark in colour. The decaying humus beneath bracken is thought to be the natural habitat for the deposition of the larva. The presence of bracken also seems an important point in favour of the fly; during a spell of cold weather or at sunset the flies are to be found on the under sides of the fronds. *H. equina* attacks both cattle and horses, on the former it is to be found congregated under the tail, along the perinaeum and sometimes near the udder. Having found a suitable spot on the skin for feeding, the fly makes one or two preliminary punctures until the haustellum comes in contact with a suitable blood vessel under the skin. Feeding continues for 7–24 minutes, and while feeding the flies are practically unaffected by external influences such as movement of the skin of the host, etc.

Melophagus ovinus Linn. The Sheep Ked; the Sheep 'Tick'

DESCRIPTION. *Adult.* Head large, flattened, slightly broader than thorax, brown, whitish beneath; eyes brown. Thorax bristly, tuberculate. Abdomen round, ferruginous, tuberculate, segments poorly differentiated. Wings absent. Halteres absent. Legs stout, yellowish brown, bristly, tarsal claws black, long and curved. Length 5–6 mm.

LIFE HISTORY. The whole life of the fly is spent on the sheep, and the flies seem unable to live for more than a few days away from the host. The egg and larval stages are passed inside the body of the female fly, the adult larva is extruded and pupates within 12 hours, the pupa remaining glued to the wool of the host. The adults hatch after 19–24 days, and the whole life cycle can be completed in 39–49 days. This fly has been shown to be the intermediate host of the blood trypanosome, *Trypanosoma melophagium* (Hoare (80)).

CONTROL. The only remedial measure seems to be the use of dips, those containing arsenic being the most successful. Carbolic acid and tobacco extracts are also used in dipping for *M. ovinus*. The treatment should be repeated two or three times at intervals of 20–25 days.

REFERENCES

DIPTERA (contd.)

Muscidae

(1) ALTSON, A. M. (1920). The life history and habits of two parasites of blow-flies. *Proc. Zool. Soc. London*, XV.

(2) BISHOPP, F. C. (1913). The stable fly (*Stomoxys calcitrans* Linn.). An important livestock pest. *Journ. Econ. Entom.* VI.

(3) GRAHAM SMITH, G. S. (1916). Observations on the habits and parasites of common flies. *Parasitology*, XI.

(4) HASEMAN, L. (1927). Controlling horn and stable flies. *Bull. Missouri Agric. Exp. Station*, No. 254.

248   DIPTERA

(5) JAMES, H. C. (1928). Cynipid parasites of Dipterous larvae. *Ann. App. Biol.* XV, No. 2.

(6) MacDOUGALL, R. S. (1909). Sheep maggot and related flies. *Trans. of the High. and Agric. Soc. of Scotland*, XXI.

(7) MACGREGOR, M. E. (1914–15). The posterior stigmata of Dipterous larvae as a diagnostic character. *Parasitology*, VII, 2.

(8) Ministry of Agriculture and Fisheries, London (1929). Advisory Leaflet No. 4.

(9) MITZMAIN, M. B. (1913). The bionomics of *Stomoxys calcitrans* L. A preliminary account. *Philippine Journ. Science*, VIII. B. No. 1. pp. 29–48.

(10) NEWSTEAD, R. (1906). On the life-history of *Stomoxys calcitrans* L. *Journ. Econ. Biol.* I.

(11) PARMAN, D. C., LAAKE, E. W. et al. (1928). Tests of blowfly baits and repellents during 1926. *U.S. Dept. Agric. Tech. Bull.* No. 80.

(12) RILEY, W. A. and JOHANNSEN, O. A. (1915). *Handbook of medical entomology*. Comstock Publishing Co., Ithaca, New York.

(13) SMIT, B. (1929). The sheep blow-flies of South Africa. *Bull. Dept. Agric. S. Africa*, No. 47. Pretoria.

Anthomyidae

(14) BAKER, A. D. (1928). The habits of onion maggot flies (*H. antiqua* Meig.). *58th Ann. Rept. Entom. Soc. Ontario*, for 1927.

(15) BAKER, A. D. and STEWART, K. E. (1927–28). The onion root maggot (*H. antiqua*) on the island of Montreal. *20th Rept. Quebec Soc. Prot. Plants*.

(16) BLUNCK, H., BREMER, H. and KAUFFMANN, O. (1929). Untersuchungen zur Lebensgeschichte und Bekämpfung der Rübenfliege (*Pegomyia hyoscyami* Pz.). *Arb. biol. Reichsanst. Land- u. Forstw.* XVII, No. 2.

(17) CAMERON, A. E. (1914). A contribution to a knowledge of the Belladonna leaf-miner *Pegomyia hyoscyami* Pz. Its life-history and biology. *Ann. App. Biol.* I.

(18) CAMERON, A. E. (1916). Some experiments on the breeding of the mangold fly (*Pegomyia hyoscyami* Pz.) and the dock fly (*P. bicolor* Wied.). *Bull. Entom. Res.* VII, No. 1.

(19) CHEVALIER, L. (1928). Biologie de la mouche de la betterave (*P. hyoscyami* Pz.). *Bull. Soc. Sci. Seine et Oise*, IX, No. 3.

(20) CORY, N. E. (1916). Notes on *Pegomyia hyoscyami* Panz. *Journ. Econ. Entom.* IX.

(21) DRUK, TWEEDE (1921). De Koolvlieg (*Chortophila brassicae* Bché). *Plantenziektenkundigen Dienst te Wageningen*, No. 8.

(22) FROST, S. W. (1924). A study of the leaf-mining Diptera of North America. *Agric. Exp. Station, Cornell Univ. Memoirs*, No. 78.

(23) GEMMILL, J. F. (1927). On the life-history and bionomics of the wheat bulb fly. *Proc. Royal Phys. Soc. Edin.* XXI, No. 3.

(24) GIBSON, A. and TREHERNE, R. C. (1916). The cabbage root maggot and its control in Canada. *Dept. Agric. Entom. Branch. Bull.* No. 12. Ottawa.

(25) GLASGOW, HUGH (1925). Control of the cabbage maggot in the seed bed. *Agric. Exp. Station, New York State Bull.* No. 512.

(26) GLASGOW, HUGH and COOK, H. T. (1929). The onion maggot situation in New York. *Journ. Econ. Entom.* XXII.

(27) KÄSTNER, A. (1929). Untersuchungen zur Zwiebelfliege (*Hylemyia antiqua* Meig.). II. Teil. Morphologie und Biologie. *Zeitschr. Morph. u. Oekol. Tiere*, XV, also III. Teil. *Zeitschr. f. Pflanzenkrank.* XXXIX, Nos. 8–9 and 10–11.

(28) KEMNER, N. A. (1925). Betflugan. *Meddelande* No. 288 *från Centralanstalten för försöksväsendet på jordbruksområdet. Entomologiska Avdelningen*, No. 47.

(29) LEACH, J. G. (1926). The relation of the seed corn maggot to the spread and development of potato blackleg in Minnesota. *Phytopathology*, XVI.

(30) MEADE, R. H. (1882). *Entom. Mon. Mag.* XIX, 148.

(31) MEADE, R. H. (1897). *A descriptive list of British Anthomyidae.* Gurney and Jackson, London.

(32) Ministry of Agriculture and Fisheries, London. Leaflet Nos. 7, 122.

(33) MORRIS, H. M. (1925). Note on the wheat bulb fly. *Bull. Entom. Res.* XV.

(34) NEEDHAM, J. G., FROST, S. W. and TOTHILL, B. H. (1928). *Leaf-mining insects.* Baillière, Tindall and Cox, London.

(35) PETHERBRIDGE, F. R. (1921). Observations on the life history of the wheat bulb fly. *Journ. Agric. Sci.* XI, Part 1.

(36) ROSTRUP, SOFIE (1924). Kornets Blomsterflue (*Hylemyia coarctata*). *Tidsskrift for Planteavl*, XXX.

(37) SAKHAROV, N. (1923). *The cause of immunity of some forms of wheat in relation to* Phorbia genitalis *Schnabl.* Entom. Dept. Saratov, Agric. Exp. Station, Saratov. (In Russian, with a summary in German.)

(38) SANDERS, J. G. (1915). *Journ. Econ. Entom.* VIII.

(39) SCHANDER, R. and GOTZE, G. (1929). Erfahrungen über das Auftreten und die Bekämpfung der Rübenfliege im Jahre 1928. *Fortsch. Landw.* IV, No. 6.

(40) SCHOENE, W. J. (1916). The cabbage root maggot, its biology and control. *New York Agric. Exp. Sta. Bull.* No. 419.

(41) SEGUY, E. (1923). *Faune de France.* Diptères Anthomyides.

(42) SEVERIN, H. H. P. and SEVERIN, H. C. (1915). Life-history, natural enemies and the poisoned bait spray as a method of control of the imported onion fly. *Journ. Econ. Entom.* VIII, No. 3.

(43) SLINGERLAND, M. V. (1894). The cabbage root maggot. *Agric. Exp. Station, Cornell Univ. Bull.* No. 78.

(44) SMITH, KENNETH M. (1922). A study of the life-history of the onion fly. *Ann. App. Biol.* IX, Nos. 3 and 4.

(45) SMITH, KENNETH M. (1925). Further experiments in the control of certain maggots attacking the roots of vegetables. *Ann. App. Biol.* XII, No. 1.

(46) SMITH, KENNETH M. (1927). A study of *Hylemyia brassicae* Bouché. The cabbage root fly. *Ann. App. Biol.* XIV, No. 3.

(47) WADSWORTH, J. T. (1915). Notes on some Hymenopterous parasites bred from the pupae of *Chortophila brassicae* Bouché, and *Acidia heraclei* L. *Ann. App. Biol.* XI.

(48) WADSWORTH, J. T. (1917). Tarred felt discs for protecting cabbages and cauliflowers from attacks of cabbage root fly. *Ann. App. Biol.* IV.

(49) WALTON, C. L. (1928). Some experiments for the control of mangold fly. *Welsh Journ. Agric.* 2.

250 DIPTERA

Psilidae

(50) GLASGOW, H. and GAINES, J. G. (1929). The carrot rust fly problem in New York. *Journ. Econ. Entom.* XXII.

(51) SMITH, KENNETH M. and GARDNER, J. C. M. (1922). *Insect pests of the Horticulturist.* Onion, carrot and celery flies. Benn Bros. London.

(52) WHITCOMB, W. D. (1929). Observations on the carrot rust fly. *Journ. Econ. Entom.* XXII.

Trypetidae

(53) FEYTAUD, J. (1914). La mouche du céleri. *Bull. Soc. Étude Vulg. Zool. Agric.* XIII. Bordeaux.

(54) FROST, S. W. (1924). A study of the leaf-mining Diptera of North America. *Agric. Exp. Station, Cornell Univ. Memoirs,* No. 78.

(55) LUNDBLAD, O. and LINDBLOM, A. (1925). Selleriflugan (*Philophylla* (*Acidia*) *heraclei* Linn.). *Meddelande* No. 283 *från Centralanstalten för försöksväsendet på jordbruksområdet. Entomologiska Avdelningen,* No. 45.

(56) Ministry of Agriculture and Fisheries, London. Leaflet No. 35.

(57) NEEDHAM, J. G., FROST, S. W. and TOTHILL, B. H. (1928). *Leaf-mining insects.* Baillière, Tindall and Cox, London.

(58) SMITH, KENNETH M. and GARDNER, J. C. M. (1922). *Insect pests of the Horticulturist.* Benn Bros. London.

Oscinidae

(59) ALDRICH, J. M. (1920). European frit fly in North America. *Journ. Agric. Res.* XVIII, No. 9.

(60) CUNLIFFE, N. (1921). Preliminary observations on the habits of *Oscinella frit. Ann. App. Biol.* VIII, No. 2.

(61) CUNLIFFE, N. (1924). Further observations on the prevalence and habits of *Oscinella frit. Ann. App. Biol.* XI, No. 1.

(62) CUNLIFFE, N. (1925). Studies on *Oscinella frit* Linn. A preliminary investigation of the recovery powers of oats when subject to injury. *Ann. App. Biol.* XII, No. 2.

(63) CUNLIFFE, N. (1925). Studies on *Oscinella frit* Linn. A note on the seasonal regularity of the maximum prevalence of the fly in the field. *Ann. App. Biol.* XII, No. 4.

(64) CUNLIFFE, N. (1928). Studies on *Oscinella frit* Linn. Observations on infestation and yield, etc. *Ann. App. Biol.* XV, No. 3.

(65) CUNLIFFE, N. and FRYER, J. C. F. (1924). *Oscinella frit* Linn. An investigation to determine how far varietal differences may influence infestation of the oat plant. *Ann. App. Biol.* XI, Nos. 3, 4.

(66) CUNLIFFE, N., FRYER, J. C. F. and GIBSON, G. W. (1925). Studies on *Oscinella frit* Linn. Correlation between stage of growth of stem and susceptibility to infestation. *Ann. App. Biol.* XII, No. 4.

(67) FREW, J. G. H. (1923). On the larval anatomy of *Chlorops taeniopus* Meig. and two related Acalyptrate Muscids. *Proc. Zool. Soc. London.*

(68) FREW, J. G. H. (1924). On *Chlorops taeniopus* Meig. *Ann. App. Biol.* XI, No. 2.

(69) FRYER, J. C. F. and COLLIN, J. E. (1924). Certain aspects of the damage to oats by the frit fly. *Ann. App. Biol.* XI, Nos. 3, 4.

(70) FULMEK, L. (1911). Zum Auftreten der Halmfliege (*Chlorops taeniopus* Meig.) in Weizen. *Osterreichische Agrar. Zeitung.* Nr. 30 vom 29.

(71) GOODEY, T. (1930). On a remarkable new nematode, *Tylenchinema oscinellae* gen. et sp. n. parasitic on the frit fly, *Oscinella frit* Linn., attacking oats. *Phil. Trans. R. Soc.* (B) CCXVIII.

(72) HEWITT, T. R. (1914). The larva and puparium of the frit fly. *Sci. Proc. Roy. Dublin Soc.* XIV, No. 23.

(73) IMMS, A. D. (1930). Observations on some parasites of *Oscinella frit* Linn. Part I. *Parasitology*, XXI, No. 4.

(74) Ministry of Agriculture and Fisheries, London. Leaflet No. 24.

(75) NOWICKI, M. (1871). Ueber die Weizenverwüsterin *Chlorops taeniopus* Meig. und die Mittel zu ihrer Bekämpfung. *Verh. zoologisch-botanischen Gesellschaft in Wien.*

(76) ROEBUCK, A. (1920). Frit fly (*Oscinis frit*) in relation to blindness in oats. *Ann. App. Biol.* VII, Nos. 2, 3.

(77) TAYLOR, T. H. (1918). *The frit fly on oats.* Univ. of Leeds and Yorks Council Agric. Educ. No. 108.

Hippoboscidae

(78) BEDFORD, G. A. H. (1926). The sheep ked (*Melophagus ovinus* Linn.). *Journ. Dept. Agric. Union S. Africa*, XII, No. 5.

(79) FREUND, L. and STOLZ, A. (1928). Beiträge zur Biologie der Schaflausfliege (*Melophagus ovinus* Linn.). *Prager Arch. Tiermed.* No. 8. Prague.

(80) HOARE, C. A. (1923). An experimental study of the sheep trypanosome and its transmission by the sheep ked. *Parasitology*, XV, No. 4.

(81) IMES, MARION (1917). The sheep tick. *U.S. Dept. Agric. Farmers' Bulletin*, No. 1798.

(82) NEWMAN, L. J. (1925). External parasites of sheep. *Journ. Dept. Agric. W. Australia*, II, No. 1.

(83) ROBERTS, J. I. (1925). On the bionomics of *Hippobosca equina. Annals Trop. Med. and Parasitology*, XIX, No. 1.

(84) SWINGLE, L. D. (1913). The life-history of the sheep tick (*Melophagus ovinus* Linn.). *Agric. Exp. Station, Univ. Wyoming, Laramie, Bull.* No. 99.

CHAPTER XIV

INSECTS AND VIRUS DISEASES OF CROPS

The virus diseases of agricultural crops are of such great economic importance and the part played by insects is so fundamental to the spread of many of these diseases that a book on agricultural entomology would be incomplete without some account of this partnership between insect and virus.

It may be well, in the first instance, to state briefly some of the outstanding characteristics of viruses. All types of organisms from man to bacteria are liable to attack by virus diseases and about two hundred have now been described affecting plants of different kinds. Most animal viruses and, so far as we know, all plant viruses are beyond the resolving power of the microscope using visible light and some are of molecular size. The virus of turnip yellow mosaic, for example, is about five times the diameter of a haemoglobin molecule. No virus has been cultivated on a cell-free medium such as agar, as have so many bacteria and fungi; a living susceptible cell is essential for virus multiplication. Most plant viruses are systemic in their hosts, all parts of the plant being invaded with the usual exception of the seed. It is a curious fact that very few plant viruses are seed-transmitted but any plant which is vegetatively propagated will, if virus-infected, give rise to virus-infected progeny. It is this fact which makes virus diseases of fundamental importance in such crops as potatoes, hops, strawberries, raspberries, dahlias, etc., all of which are vegetatively propagated.

Several plant viruses have been isolated in pure crystalline form and shown to be nucleo-proteins. One of these is the virus causing turnip yellow mosaic [1] which is described in a later paragraph.

Not all plant viruses are insect-transmitted but in this chapter attention is confined to those which are spread in this manner. The most important group of insects concerned with the spread of plant viruses is the Hemiptera, especially leaf-hoppers, aphides and whitefly, but in Great Britain the only insects of this order known to be vectors of viruses are the aphides. The mechanics of inoculation of a healthy plant by a Hemipterous insect are thought to be briefly as follows. The insect, when feeding upon the plant, inserts the mandibles and maxillae, not the labium, into the tissue (Fig. 77) and draws the sap up one of the two channels which are formed by the close apposition to each other of the mandibles and maxillae,

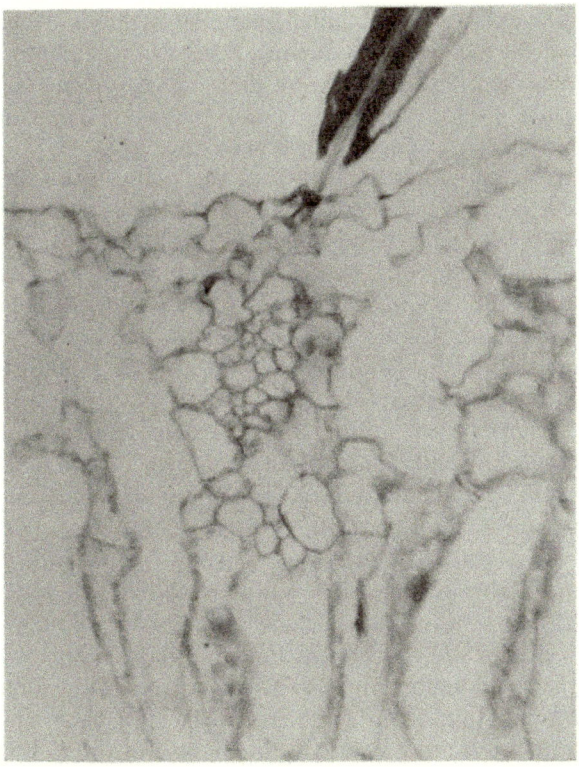

Fig. 77. Photomicrograph showing the method of feeding of the aphis *M. persicae*, an important insect vector of potato virus diseases. In the section of a potato leaf shown, the stylets of the aphis may be seen in the act of penetrating a stoma; the objective is the phloem which lies just below. × 650.

and pumps saliva down the other channel by means of the pharyngeal pump situated in the head. If the plant on which the insect is feeding is virus-infected, then the virus-containing sap mixes with the insect's saliva which thus becomes contaminated with virus. Then on migrating to, and feeding upon, a healthy susceptible plant, some of the virus is injected with the saliva into the new host which in turn develops the disease.

The relationship of viruses with insects of this type is, broadly speaking, of two kinds: those viruses, called *non-persistent*, which are rapidly lost by the insect unless it has recourse to a fresh source of virus, and those viruses, called *persistent*, which are retained for long periods by the insect, frequently for the rest of its life, without having access to a fresh source of virus. Potato virus Y, causing severe mosaic, is an example of the non-persistent virus and potato leaf-roll virus is an example of the persistent type.

Whilst it is undoubtedly true that the majority of the insect-transmitted plant viruses are carried by insects of the order Hemiptera, it has recently been shown that biting insects can also act as vectors. The flea beetles, *Phyllotreta* spp., particularly *P. cruciferae* and *P. undulata*, transmit the virus disease of turnips, already mentioned, known as turnip yellow mosaic (2). This is a fact of some significance and increases the already great importance of the turnip flea beetles.

From the agricultural point of view the virus diseases of potatoes, sugar beet, leguminous and cruciferous crops are the most important and these will be briefly described together with their insect vectors and methods of control.

Potato virus diseases. There are several viruses attacking the potato crop in the British Isles, not all of which are insect-transmitted; attention, however, is here confined to those which are carried by insects and the first is known as potato *leaf-roll*. As its name implies, the main symptom of this disease is a rolling of the leaves. The first sign of infection is usually a pallor at the base of the youngest leaves and this is followed by inward rolling of the leaf edges, beginning at the bases of the leaves. In a primary infection only the central leaves are rolled but when the secondary symptoms have developed, all the leaves show rolling to a greater or less degree, the plant is stunted and the leaves are thickened and discoloured. (Fig. 78.) The reduction in yield caused by leaf-roll may be as much as 90 per cent. of the crop. The leaf-roll virus cannot be transmitted by mechanical inoculation and is entirely dependent upon the aphis for its spread in the field; no other type of insect can carry the virus. The most important species is *Myzus persicae* Sulz. (see p. 50), but there are one or two other species which act as subsidiary vectors. These are *M. convolvuli* Kalt. (=*pseudosolani* Theob.) (see p. 54) and *M. ornatus* Laing. As already mentioned, the leaf-roll virus is of the persistent type and is retained by the aphis for long periods without the necessity for the insect to have recourse to a fresh source of virus.

Fig. 78. Potato leaf-roll: plants experimentally infected by
means of the aphis *Myzus persicae*.

Another important potato virus, also aphis-transmitted, is that known as potato virus Y, so-called to differentiate it from potato virus X which, so far as is known, is not insect-transmitted. Virus Y gives rise to the disease in potatoes, known as *severe mosaic* or *leaf-drop streak*. (Fig. 79.) The first sign of infection is the appearance on the undersides of the leaves of small dark necrotic streaks running along the veins. These streaks increase in size and coalesce, the leaves finally shrivelling and hanging down, but remaining attached to the plant in the manner shown in Fig. 79. The top of the plant does not shrivel but usually shows a mottling.

There is one other aphis-transmitted virus affecting potatoes which should be mentioned; this is known as potato virus A and is similar in many respects to potato virus Y. By itself virus A is not of great importance; its significance lies in the fact that together

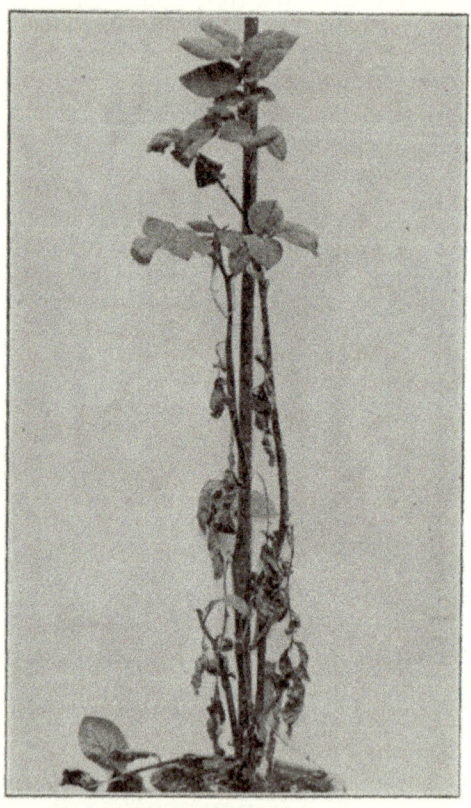

Fig. 79. President potato plant infected with leaf-drop streak or severe mosaic caused by the aphis-transmitted potato virus Y.

with virus X, it gives rise to the familiar virus disease of potatoes known as *crinkle*. The chief symptoms of this disease are a puckering and downward curving of the leaves which are crinkled at the edges. Diffused yellowish areas occur all over the foliage and this is followed by the appearance of rusty brown spots which develop at the tips of the leaves. Affected plants are bushy and stunted and usually stand out from neighbouring plants by reason of their paler green colour. Since potato virus X is not aphis-borne, the crinkle disease is not transmitted as such but develops in plants, already infected with virus X, to which virus A is carried by the aphides.

Control of potato virus diseases. We are only concerned here with the control of the insect-borne potato virus diseases, so that the problem resolves itself into the question of the elimination or avoidance of the potato-feeding aphides. In practice it is the latter procedure which is adopted. It is of course essential that the 'seed' should be as virus-free as possible so that the aphides, if present, will not have an abundant source of virus at hand, and it is for this reason that the seed potatoes are obtained by the English farmer from Scotland and Ireland. Indeed the seed-potato growing areas of Scotland and Ireland are a practical illustration on a large scale of the control of potato viruses by the avoidance of the chief aphis vector, *M. persicae*. In these districts the climate is unsuitable for this insect which will not fly if there is a prevailing wind blowing, of more than about 4 m.p.h.; if the temperature is lower than about 65° F. and if relative humidity is more than 75 per cent. These conditions are all fulfilled in the seed-growing districts and in consequence the aphis is very scarce in those areas. However, much can be done even in England to avoid the aphis, if some attention is paid to its life-history. For example, it is known that *M. persicae* passes the winter in various ways; as a winter egg on peach trees; in glasshouses where it breeds on many different kinds of plants all the year round and, in mild winters, it can over-winter out-of-doors on brassicas such as Brussels sprouts. It is best, therefore, to avoid whenever possible, growing potatoes in areas where the aphis can find these facilities for over-wintering.

Finally, since aphides are liable to feed on the sprouts of tubers in the sprouting trays and since viruses can be disseminated in this way, it is advisable to fumigate the potato store with nicotine at regular intervals.

Virus diseases of sugar-beet and mangolds. In the British Isles there are two viruses affecting sugar-beet and mangolds, those of *mosaic* and *virus yellows*, and the second is the more important of the two. Both viruses are transmitted by the aphides *M. persicae* Sulz. and *Aphis fabae* and not by any other insect. The mosaic virus is of the non-persistent, and the yellow virus of the persistent, type. In the field, spread of these viruses is entirely dependent upon aphides.

The first sign of infection in sugar-beet mosaic is a 'clearing' or picking out of the veins of the youngest leaves. This is followed by

Fig. 80.   Leaf of sugar-beet plant infected with sugar-beet mosaic.

a light and dark green mottle which develops on the next youngest leaves. In sugar-beet yellows there is less preliminary clearing of the veins and no mosaic mottling. The outer and middle leaves become thickened, yellow and brittle. This yellowing, which usually begins at the tips and spreads downwards, varies from orange-yellow to scarlet or russet-brown. Later these yellowed parts turn brown and wither. On walking through a field of beet heavily infected with yellows, there is a pronounced crackling of the brittle leaves under foot. The chief damage caused by this

Fig. 81. Leaf of sugar-beet plant infected with virus yellows.

disease is the reduction in the sugar content and the earlier the crop is infected the greater the reduction. A crop of sugar-beet heavily infected in July may lose up to 50 per cent. of its sugar (3).

Neither virus is seed-transmitted so that the crop will start free of infection which is always heaviest in the seed-growing districts. It is important to eliminate sources of infection such as volunteer beets and mangolds and to clear the mangold clamps by the end of March. These clamps act not only as a source of virus infection but also of the transmitting aphides. If possible the root crop should not be grown close to the seed crop. Early sowing of the commercial root crop is recommended with late sowing (second week in August) of the stecklings. Some degree of control may be obtained by fortnightly treatments of the steckling beds with an aphicide such as nicotine.

Virus diseases of leguminous crops. There are at least six viruses which attack peas, beans and clover in the British Isles, but not all of them have been properly studied. The chief insect vector seems to be the pea and clover aphis, *Macrosiphum pisi* but *Myzus persicae* and *Aphis fabae* are also concerned in the spread of some of these viruses. In pea mosaic the first sign of infection is a clearing or yellowing of the veins of the youngest leaves. This is followed by a mottling which is sometimes a very pronounced green and yellow mosaic, accompanied by stunting and dwarfing of the plant. This virus also affects the broad bean (*Vicia faba*) on which it produces a rather bright yellow and green mottle.

Pea mosaic virus is also the commonest and most important virus which attacks clovers. The following are all susceptible, trefoil (*Medicago lupulina*) and the wild black medick, crimson clover (*Trifolium incarnatum*), red clover (*T. pratense*) and alsike (*T. hybridum*). On all these plants the virus produces a mosaic mottling of varying intensity. The susceptibility of clovers to the pea mosaic virus has an added significance in that they are the winter host of the pea aphis (*Macrosiphum pisi*) and so act as a reservoir of virus from which infection is carried to the pea crop by the migrating aphides in spring.

Virus diseases of cruciferous crops. As in the case of the viruses affecting leguminous crops, more work is required on the virus diseases of Cruciferae. There are, however, three viruses which have been studied in some detail and are of considerable importance. All three are insect-borne. The first of these is known as the cabbage black ringspot virus and it attacks cabbages, cauliflowers, Brussels sprouts, turnips and swedes. It also affects many ornamental plants belonging to the Cruciferae, especially stocks and wallflowers and causes a 'break' or colour change in the flowers. As its name implies the main symptom on members of the cabbage family is the development of numerous black rings and spots, followed by blackening of the veins and midribs. On swedes and turnips the disease takes the form of a mosaic mottling with considerable dwarfing and stunting in the case of the turnip. The chief aphis

Fig. 82. Clover mosaic: the virus is brought from infected peas by the aphis *Macrosiphum pisi*.

vectors of this virus are *Myzus persicae* and the mealy cabbage aphis, *Brevicoryne brassicae*.

In *cauliflower mosaic* the first sign of infection is a clearing of the veins of the youngest leaves, a point of difference from the black ringspot virus. This is followed by a characteristic banding of the veins with a deeper green colour. The effect of this disease on cauliflower is severe and the curd is so reduced as to be useless for market. On turnips the virus produces a pronounced rosetting of

Fig. 83. Cabbage black ringspot; leaf of young cabbage plant experimentally infected by means of the aphis *Myzus persicae*.

the crown; the central leaves are curled inwards and the veins stand out with a yellow colour. The whole plant is greatly stunted. The same two aphides which transmit the cabbage black ringspot virus are the vectors of the cauliflower mosaic virus.

Control of cabbage black ringspot and cauliflower mosaic. In many parts of the British Isles, especially in the broccoli-growing districts, these two diseases are of great importance. In western Washington, also in the U.S.A., over 60 per cent. of the plants of the entire cabbage seed acreage for 1943 and 1944 harvests were infected;

Fig. 84. Leaf of turnip infected with yellow mosaic; this virus is transmitted by the turnip flea beetle, *Phyllotreta* spp.

this resulted in a severe reduction in seed yield. Since it is difficult to control the cabbage aphis, attempts were made to avoid infection during the period when infected aphides were migratory. This was done by isolation of the plant beds from the diseased cabbage fields. In seed fields of the 1944–5 crop, planted from isolated plant beds, the amount of mosaic 7–8 months after transplanting was only 5 per cent. In the 1945–6 crop, over 90 per cent. of the plants were grown in well-isolated areas and the average infection of the entire acreage 8 weeks after transplanting was only 2·4 per cent. (4). It seems probable that in this country also, most of the infection takes place among the young plants prior to setting out, so that some system of isolation or screening of the seedling plants appears to offer the best chances of successful control.

Turnip yellow mosaic. This disease is chiefly of interest because of its association with a new type of insect vector. The symptoms consist of an unusually bright mottling of yellow and green, so bright as to resemble a variegation rather than a mosaic (Fig. 84). Affected plants are smaller than normal and make poor growth because of the lack of chlorophyll.

The virus attacks turnips and swedes; it also affects radishes, kohlrabi and the weed Shepherd's Purse; cabbages and cauliflowers do not appear to be susceptible. The insect vector is a flea beetle, *Phyllotreta* spp., and more than one species are able to transmit the virus. *P. cruciferae* and *P. undulata* seem to be the commonest vectors (see pp. 117–8).

This is the first record of a biting insect transmitting a virus in the British Isles and the mechanism of infection is not yet clear. It does not, however, appear to be a merely mechanical process since the beetles can retain the virus for 3 days and infect a number of plants in succession. Possibly infection may take place by regurgitation, since many beetles use this method as an aid to digestion, salivary glands being apparently absent.

REFERENCES

(1) MARKHAM, R. and SMITH, KENNETH, M. (1946). A new crystalline plant virus. *Nature, Lond.*, CLVII, 300.
(2) SMITH, KENNETH M. and MARKHAM, R. (1946). An insect vector of the turnip yellow mosaic. *Nature, Lond.*, CLVIII, 417.
(3) Sugar-beet yellows. (1946). Advisory Leaflet No. 323. Ministry of Agriculture and Fisheries.
(4) POUND, G. S. (1946). Control of virus diseases of cabbage seed plants in Western Washington by plant bed isolation. *Phytopath.* XXXVI, 1035–9.

BOOKS ON PLANT VIRUS DISEASES

BAWDEN, F. C. (1943). Plant viruses and virus diseases. 2nd edn. *Chron. Botanica Co. Waltham, Mass.* Wm. Dawson and Sons, London, W. 1.
SMITH, KENNETH M. (1947). *Plant Viruses.* Methuen's Biol. Mon. 2nd edn.
SMITH, KENNETH M. (1945). *Virus diseases of farm and garden crops.* Littlebury and Co., Worcester.

APPENDIX I

CHARACTERISTIC SYMPTOMS OF INSECT ATTACK

Insects which attack agricultural crops often produce injuries which are characteristic of the species concerned, and the identity of the insect can sometimes be determined from the nature of the symptoms exhibited by the affected plant. To assist the student in his observations in the field upon damaged crops, the following appendix has been prepared in which the main agricultural crops are listed, together with the more important insect pests and the outstanding symptoms of their attack. In each case the reader is referred to the page in the text dealing with the insect in question.

Crop	Chief symptoms of attack	Insect attacking
CRUCIFERAE		
Cabbage, cauliflower, etc.	Leaves skeletonised	Cabbage whites, *Pieris brassicae, P. rapae* (pp. 68 and 70) Cabbage moth, *Barathra brassicae* (p. 77)
Cabbage and cauliflower	Blackened tunnels in the stems and midribs of the leaves; larvae usually present in the borings	Cabbage stem flea beetle, *Psylliodes chrysocephala* (p. 121; Fig. 30)
Swedes	Stalks swollen and bent sharply inwards across the top of the plant compressing the terminal bud, leaves crumpled; also the 'many neck' condition, where numbers of heads are formed, is sometimes present	Swede midge, *Contarinia nasturtii* (p. 175; Fig. 48)
Turnips	Hollowing out of the roots	Turnip moth, *Euxoa segetum* (p. 72)
Turnips	Leaves skeletonised; numbers of black larvae present	Turnip sawfly, *Athalia spinarum* (p. 149)
Turnips, cabbage, cauliflower, etc.	Marble-like galls or round swellings on the roots, containing a larva or larval cell	Turnip gall weevil, *Ceuthorrhynchus pleurostigma* (p. 138; Fig. 36)
Turnips, cabbage, cauliflower, etc.	Plant limp and flaccid, leaves at first bluish, later turning yellow; a number of maggots in the roots	Cabbage root fly, *Hylemyia brassicae* (p. 218)
Various Cruciferae	Hypocotyl of developing seed destroyed below soil level; round holes eaten in the leaves of plant	Flea beetles, *Phyllotreta* spp. (p. 115)

Crop	Chief symptoms of attack	Insect attacking
Various Cruciferae: seed pods	Pod discoloured and deformed; larva feeding on the unripe seeds	Turnip seed weevil, *Ceuthorrhynchus assimilis* (p. 141)
Seedling crops of many kinds but often Brassicae	Stem severed or eaten into below ground	Leatherjackets, *Tipula* spp. (p. 156)
Seedling crops of many kinds but often Brassicae	Stem of plant severed at or just below ground level	Cutworms, *Noctuidae* (p. 72)

LEGUMINOSAE

Peas	Pods undersized and with a characteristic silvery appearance	Pea thrips, *Frankliniella robusta* (p. 24; Fig. 3)
Peas	Seeds with a single cell containing larva or beetle	Pea beetle, *Bruchus pisi* (p. 131)
Peas	U-shaped notches eaten in the leaf edges	Pea weevil, *Sitona* spp. (p. 135)
Peas	Flowers shrivelled and the base of the sepals swollen, petals crinkled and deformed, pods contain large numbers of jumping larvae	Pea midge, *Contarinia pisi* (p. 182; Fig. 49)
Beans	Stems hollowed, containing yellow larvae	Wireworms, *Agriotes* spp. (p. 102)
Broad beans	Seeds with a number of cells containing larvae or beetles	Bean beetles, *Bruchus rufimanus* and *B. affinis* (pp. 131–2)
Clover	Small larvae feeding in the green flower heads	Clover weevils, *Apion* spp. (p. 143)
Clover	Leaflets unopened, brown and swollen, forming a gall in which are reddish yellow larvae	Clover leaf midge, *Dasyneura trifolii* (p. 180)

UMBELLIFERAE

Carrots	Tops suffused with coppery red, later turning yellow. Root with rusty red borings in the cortex, containing larvae	Carrot fly, *Psila rosae* (p. 228)
Celery	A large blotchy type of blister in the leaves which later become brown and curled; a number of larvae in the blisters	Celery fly, *Acidia heraclei* (p. 231)
Celery	Notches eaten in the leaf edges	*Phaedon tumidulus* (p. 125)

SOLANACEAE

Potatoes	Plant wilting, stem completely hollowed out	Rosy rustic moth, *Gortyna micacea* (p. 81)
Potatoes	Tubers riddled with holes	Wireworms, *Agriotes* spp. (p. 102)
Potatoes	Large numbers of brown spots on leaves and shoots, sometimes with a 'shot-hole' effect; shoots often killed	Potato capsid bugs, *Calocoris norvegicus*, *Lygus pabulinus* and *Lygus pratensis* (p. 31). Fig. 4

Crop	Chief symptoms of attack	Insect attacking
Potatoes	Large numbers of white spots on the leaves, coalescing in a bad attack to form a bleached area	Potato leaf hoppers, *Eupteryx auratus* and *Chlorita viridula* (pp. 39–40)
Potatoes	Round holes in leaves; roots attacked by small larvae	Potato flea beetle, *Psylliodes affinis* (p. 123)

CHENOPODIACEAE

Mangolds	A constriction of the root at or just *above* ground level, occasionally accompanied by round holes in the leaves	Spring-tails, *Bourletiella hortensis* (p. 20)
Mangolds	A constriction of the root *below* soil level	Pigmy mangold beetle, *Atomaria linearis* (p. 94)
Mangolds	Blisters in the leaves containing Dipterous larvae	Mangold fly, *Pegomyia hyoscyami* (p. 224)

URTICACEAE

Hop	Young leaves spun together, a round hole in the shoot	Rosy rustic moth, *Gortyna micacea* (p. 81)
Hop	Bracts of hop cone riddled with small holes	Hop cone flea beetle, *Psylliodes attenuata* (p. 124)
Hop	Shoots punctured and often killed; usually a dense growth of lateral shoots present	Needle-nosed hop bug, *Calocoris fulvomaculatus* (p. 36)

AMENTACEAE

Willow	Bean-shaped green or red galls on the leaves	Willow gall sawfly, *Pontania proxima* (p. 154)
Willow	'Shot-holes' in the bark	Shot-hole borer, *Rhabdophaga saliciperda* (p. 174)
Willow	A shapeless 'rosette' or 'button' gall containing a number of larvae	Button-top gall midge, *Rhabdophaga heterobia* (p. 172)
Willow	A rosette gall, less shapeless than the foregoing, containing a single larva	*Rhabdophaga rosaria* (p. 173)

LILIACEAE

Onions	Injury at collar, with premature seeding	Wireworms, *Agriotes* spp. (p. 102)
Onions	Leaves soft and flaccid, later turning yellow and wilting, finally lying prone; a number of maggots present in root	Onion fly, *Hylemyia antiqua* (p. 214)

GRAMINACEAE

Oats, barley, etc.	Ear with glumes whitish and paler than the rest of the ear, occasionally accompanied by blindness	Corn thrips, *Limothrips cerealium* (p. 28)

Crop	Chief symptoms of attack	Insect attacking
Wheat	In *spring*, a widening of the sheath in the last mature leaf and the yellowing of the youngest protruding leaf; caterpillar present in the stem. In *late summer*, wheat grains hollowed out, holes showing in the grains	Rustic shoulder **knot**, *Hadena basilinea* (**p.** 78)
Wheat, barley	White ears standing erect as compared with the normal growing corn, later the stems are severed; a yellowish larva present at the bottom of the stem	Wheat stem **sawfly**, *Cephus pygmaeus* (p. 151). (Fig. 39)
Cereals	Young plants attacked at the interval between the seed and the bulb; the stem often bitten through, causing the whole plant to wither, usually a number of consecutive plants in a row are killed	Wireworms, *Agriotes* spp. (p. 102)
Cereals	Foliage of young plants in spring eaten in longitudinal strips	*Lema melanopa* (p. 110). (Fig. 26)
Cereals	Root and stem below ground eaten away, or stem and blade cut off above ground	Leatherjackets, *Tipula* spp. (p. 156)
Wheat	Crop beginning to fail in June or July; bending or elbowing of the straw; whitish larvae or 'flaxseed' puparia present in the stem near a knot	Hessian fly, *Mayetiola destructor* (p. 164). (Fig. 44)
Wheat	Ears with a bluish colour at the base of the glume, small *golden yellow* larvae feeding among the anthers and pollen at the top of the kernel	Wheat midge, *Contarinia tritici* (p. 167). (Fig. 46 B)
Wheat	Small *carrot red* larvae present on the face of the kernel itself	Wheat midge, *Sitodiplosis mosellana* (p. 170). (Fig. 46 A)
Wheat	Failure of young plants in April or May, the central leaf turns yellow, later it flags and dies. There is a discoloration just above the root, where there is a single whitish maggot	Wheat bulb fly, *Hylemyia coarctata* (p. 211). (Fig. 59)
Wheat and barley	In early spring, plants show stunting of the attacked shoot, leaves broad and abnormally green, there appears to be no stem and the central leaves are spirally twisted. In June, the ears are *swollen* and unable to emerge from the spirally twisted sheathing leaves. There is a track eaten along one side of the stem from the uppermost knot into the ear.	Gout fly, *Chlorops taeniopus* (p. 235). (Fig. 73 A, B, C)

Crop	Chief symptoms of attack	Insect attacking
Oats	In *spring*, the young oat plant shows destruction of the central shoot which turns first pale green then greenish yellow and finally withers; numbers of tillers produced giving rise to a short stunted plant. In *July and August*, developing ears are infested with small larvae which destroy the grain	Frit fly, *Oscinella frit* (p. 238). (Fig. 75)
Oats	Base of plant swollen and 'tulip-rooted', leaves twisted and pale in colour, whole plant stunted. Differs from attack by frit fly in that the central shoot is not eaten away, and there are no larvae present	The stem eelworm, *Tylenchus dipsaci*
Pasture	Presence of large numbers of 'processionary' caterpillars	Antler moth, *Charaeas graminis* (p. 82)
Grass land	Roots of grass destroyed; patches of ground with a soft spongy texture	Garden chafer, *Phyllopertha horticola* (p. 98)

APPENDIX II

COMMON FARM WEEDS IN RELATION TO INSECT PESTS

In the following appendix is given a list of the common farm weeds, which act as alternate hosts for many insect pests of agricultural crops, together with the more important insects which feed upon them. Weeds of farm land, owing to possible changes in the feeding habits of the insects associated with them, acquire an added importance. On the continent wild species of *Atriplex*, for example, are regarded as a source of special danger to the sugar-beet crop owing to the transfer of insect species from this weed to the beet.

Weed	Insects
CRUCIFERAE	
Charlock, *Sinapis arvensis*	Diamond-back moth, *Plutella maculipennis*
	Turnip sawfly, *Athalia colibri*
	*Wireworms, *Agriotes* spp.
	Flea beetles, *Phyllotreta* spp.
	Mustard beetle, *Phaedon cochleariae*
	Turnip seed weevil, *Ceuthorrhynchus assimilis*
	Cabbage aphis, *Brevicoryne brassicae*
	Swede midge, *Contarinia nasturtii*
Hedge mustard, *Sisymbrium officinale*	Diamond-back moth, *P. maculipennis*
	Turnip sawfly, *A. colibri*
Winter cress, *Barbarea vulgaris*	Turnip sawfly, *A. colibri*
	Flea beetles, *Phyllotreta* spp.
Large bitter cress, *Cardamine amara*	Mustard beetle, *Phaedon cochleariae*
Bitter cress, *Cardamine hirsuta*	Large white butterfly, *Pieris brassicae*
Field cress, *Isatis tinctoria*	Cabbage aphis, *Brevicoryne brassicae*
Marsh watercress, *Nasturtium palustre*	Swede midge, *Contarinia nasturtii*
Creeping water cress, *N. silvestre*	Swede midge, *Contarinia nasturtii*
Shepherd's purse, *Capsella bursa-pastoris*	Bean aphis, *Aphis fabae*
	Cabbage aphis, *Brevicoryne brassicae*
Alyssum spp.	Small white butterfly, *Pieris rapae*
Wild radish, *Raphanus sativus*	Flea beetles, *Phyllotreta* spp.
	Swede midge, *C. nasturtii*
Treacle mustard, *Erysimum cheiranthoides*	Flea beetles, *Phyllotreta* spp.
	Cabbage aphis, *B. brassicae*
Field Brassica, *Brassica campestris*	Flea beetles, *Phyllotreta* spp.
Wall Brassica, *Diplotaxis tenuifolia*	Cabbage aphis, *B. brassicae*
Camelina spp.	Flea beetles, *Phyllotreta* spp.

* Denotes root feeders.

Weed	Insects
ONAGRACEAE Willow herb, *Epilobium* spp.	Potato and rose aphis, *Macrosiphum gei*
UMBELLIFERAE Cow parsnip, *Heracleum sphondylium*	Celery fly, *Acidia heraclei*
Wild angelica, *Angelica sylvestris*	Celery fly, *A. heraclei*
COMPOSITAE Knapweed, *Centaurea nigra*	Pea thrips, *Frankliniella robusta*
Dandelion, *Taraxacum dens-leonis*	*Ghost moth, *Hepialus humuli* Cabbage white fly, *Aleurodes brassicae* Onion fly, *Hylemyia antiqua.* (Adult fly feeds on the flowers.)
Groundsel, *Senecio vulgaris*	Springtails, *Bourletiella hortensis*
Sow thistle, *Sonchus oleraceus*	Capsid bug, *Lygus pabulinus* Potato and rose aphis, *Macrosiphum gei*
Scotch thistle, *Onopordon acanthium*	Mangold fly, *Pegomyia hyoscyami*
Thistles, *Carduus* spp.	Bean aphis, *Aphis fabae*
CONVOLVULACEAE Lesser bindweed, *Convolvulus arvensis*	Capsid bug, *L. pabulinus*
Larger bindweed, *C. sepium*	Capsid bug, *L. pabulinus*
SOLANACEAE Bittersweet, *Solanum dulcamara*	Potato flea beetle, *Psylliodes affinis* Capsid bug, *L. pabulinus* Potato aphis, *Myzus persicae*
Black nightshade, *Solanum nigrum*	Capsid bug, *Lygus pratensis* Capsid bug, *L. pabulinus* Potato aphis, *M. persicae*
Deadly nightshade, *Atropa belladonna*	Mangold fly, *Pegomyia hyoscyami*
Henbane, *Hyoscyamus niger*	Mangold fly, *P. hyoscyami*
SCROPHULARIACEAE Purple foxglove, *Digitalis purpurea*	Potato aphis, *Myzus pseudosolani*
CHENOPODIACEAE Orache, *Atriplex* spp.	Mangold fly, *P. hyoscyami*
Nettle-leaved goosefoot, *Chenopodium murale*	Mangold fly, *P. hyoscyami*
Goosefoot, *Chenopodium album*	Springtails, *Bourletiella hortensis* Green tortoise beetle, *Cassida vittata* Beet sylphid, *Blitophaga opaca* Potato and rose aphis, *Macrosiphum gei* Bean aphis, *Aphis fabae* Mangold fly, *P. hyoscyami*

* Denotes root feeders.

Weed	Insects
POLYGONACEAE	
Curled dock, *Rumex crispus* and *Rumex* spp.	Bright-line brown-eye, *Polia oleracea* Rosy rustic moth, *Gortyna micacea* *Ghost moth, *Hepialus humuli* Tooth-legged flea beetle, *Plectroscelis concinna* *Wireworms, *Agriotes* spp. Capsid bug, *L. pabulinus* Bean aphis, *Aphis fabae*
URTICACEAE	
Nettle, *Urtica dioica*	Bright-line brown-eye, *P. oleracea* Hop cone flea beetle, *Psylliodes attenuata* Capsid bugs: *Lygus pabulinus, Calocoris norvegicus, L. pratensis* Potato leaf-hopper, *Eupteryx auratus* Potato and rose aphis, *Macrosiphum gei* Potato aphis, *Myzus persicae* Hop-damson aphis, *Phorodon humuli*
GRAMINACEAE	
Cocksfoot, *Dactylis glomerata*	Springtails, *Sminthurus viridis* *Lema melanopa* Grain aphis, *Macrosiphum granarium* *Summer chafer, *Amphimallus solstitialis* *Wireworms, *Agriotes* spp. Corn thrips, *Limothrips cerealium*
Couch grass, *Agropyron repens*	Rosy rustic moth, *Gortyna micacea* *Wireworms, *Agriotes* spp. *Lema melanopa* Hessian fly, *Mayetiola destructor* Wheat midge, *Contarinia tritici* Wheat bulb fly, *Hylemyia coarctata* Gout fly, *Chlorops taeniopus* Frit fly, *Oscinella frit*
Rye grasses, *Lolium* spp.	Springtails, *S. viridis* *Turnip moth, *Euxoa segetum* Frit fly, *Oscinella frit* *Wireworm, *Athous haemorrhoidalis* *Yellow underwing moth, *Tryphaena pronuba* *Leatherjackets, *Tipula oleracea* Corn thrips, *Limothrips cerealium* *Summer chafer, *Amphimallus solstitialis* *Ghost moth, *Hepialus humuli*
Timothy grass, *Phleum pratense*	Springtails, *S. viridis* *Lema melanopa*
Canary grass, *Phleum canariensis*	*Lema melanopa*
Meadow foxtail, *Alopecurus pratensis*	Grain aphis, *Macrosiphum granarium*

* Denotes root feeders.

Weed	Insects
Sheeps' fescue, *Festuca ovina-duriuscula* and *Festuca* spp.	Grain aphis, *M. granarium* *Myzus festucae* Frit fly, *Oscinella frit*
Wild oat, *Avena fatua*	Grain aphis, *M. granarium* *Myzus festucae*
Tall oat grass, *Arrhenatherum avenaceum*	*Wireworms, *Agriotes* spp. Frit fly, *Oscinella frit* *Leatherjackets, *Tipula oleracea* Grain aphis, *M. granarium* *Lema melanopa*
Meadow grasses, *Poa* spp.	Frit fly, *Oscinella frit* Grain aphis, *M. granarium* Noctuidae, *Hadena basilinea* *Cockchafer, *Melolontha melolontha* *Yellow underwing moth, *Tryphaena pronuba* *Leatherjackets, *Tipula oleracea*
Common holcus, *Holcus lanatus* and *Holcus* spp.	Frit fly, *O. frit* Grain aphis, *M. granarium* Corn thrips, *Limothrips cerealium*
Brome, *Bromus* spp.	*Myzus festucae* Grain aphis, *M. granarium*
Bent grass, *Agrostis vulgaris*	Antler moth, *Charaeas graminis*

* Denotes root feeders.

LITERATURE ON WEEDS

BRENCHLEY, W. E. (1920). *Weeds of Farm Land.* Longmans, Green and Co. London.

LONG, H. C. (1928). *Weeds of Arable Land.* Ministry of Agriculture and Fisheries, Miscellaneous Publications, No. 61.

LONG, H. C. (1930–31). Weeds of Grass Land. *Jour. Min. Agric.* XXXVII, Nos. 9 & 10.

MILES, H. W. (1921). Observations on the insects of grasses and their relation to cultivated crops. *Ann. App. Biol.* VIII.

MORSE, R. and PALMER, R. (1925). *British Weeds; their identification and control.* Benn Bros. London.

INDEX OF AUTHORS

INDEX OF PARASITES AND PREDATORS

COLEOPTERA

HYMENOPTERA

HYMENOPTERA (*contd.*)

HYMENOPTERA (contd.)

HYMENOPTERA (*contd.*)

GENERAL INDEX

For EU product safety concerns, contact us at Calle de José Abascal, 56–1°,
28003 Madrid, Spain or eugpsr@cambridge.org.

www.ingramcontent.com/pod-product-compliance
Ingram Content Group UK Ltd.
Pitfield, Milton Keynes, MK11 3LW, UK
UKHW010852090126
466816UK00011B/175